A Portrait of Assisted Reproduction in Mexico

Sandra P. González-Santos

A Portrait of Assisted Reproduction in Mexico

Scientific, Political, and Cultural Interactions

Sandra P. González-Santos
Facultad de Bioética
Universidad Anáhuac
Mexico City, Mexico

ISBN 978-3-030-23040-1   ISBN 978-3-030-23041-8   (eBook)
https://doi.org/10.1007/978-3-030-23041-8

Cover illustration: S consultores en diseño

This Palgrave Macmillan imprint is published by the registered company Springer Nature Switzerland AG
The registered company address is: Gewerbestrasse 11, 6330 Cham, Switzerland

An Apology and a Long Thank You

Books are written by people, by people with lives that go beyond the book, lives that they share with other people. And books, or I should say the production of books, disrupt these lives. If I may be blunt, writing a book shares some similarities with raising a misbehaved unruly child who is demanding attention 24/7, and the author is like a very lenient parent who gives in to every little whim. In other words, authors writing books can be quite annoying. So, I want to apologise to those people this particular author-book annoyed and thank them for their patience and for the many long conversations they gifted me with. Because books are collective endeavours, they are summaries of long conversations, of books read, of music, of life experiences, surprises, frustrations, hopes, and imaginaries. When I found myself at the end of the process of writing, I become acutely aware of all those moments, people, places, books, and feelings. I was able to see how they had taken me to the point of writing, how they had inspired me to say something, how they had helped me to do it in the best way I could, and how they had accompanied me along the way. I want to recognise and thank each one of them.

"AR specialists" and "AR users" whom I met at the many clinics I have visited since 2006, at the different conferences and events, I thank you all for your trust, faith, and interest in my project. It goes beyond saying that this book would have not been possible without you.

Adam Hedgecoe, Nicky Priaulx, and Martin Weinel, without your some-times subtle other times not so subtle nudges I would have never had the nerve to embark on this task.

Rebecca Dimond and Niel Stephens not only have you been great friends, but you have also been my teachers. Without the many hours we have spent chatting, reading, discussing, writing, and rewriting, I would have never developed the skills to pull this off.

Stevienna de Saille and Ayo Wahlberg, the many conversations we have had across the years gave your books a special voice and made them my private confidants, my "go-to" places when I was lost in my path. Both for the conversations and for your books, thank you.

Beth Reddy, Andrew Konove, and Roger Magazine, I thank you for tak-ing the time to read my drafts, for listening to my stories, but above all, for your words of encouragement together with your instructions to go back to the desk and try again.

Sandra Harding, thank you for reading my drafts with such detail and taking the time to correct them, but mostly thank you for your support and encouragement.

Sarah Franklin (you were the first to encourage me to publish) and Marcia Inhorn, I thank you for giving me the opportunity to be part of this amazing group of scholars (my literature review!). The many hours of collective work and conversation you have gifted us with gave this book a purpose. I particularly want to thank Charis Thompson for her insightful comments that helped give this book the cohesion and unity.

No book can be a book without an editor, thank you Josh, particularly with the supportive words when I was about to lose it.

I also want to thank Dr. Cabrera and the Bioethics Faculty of the Universidad Anáhuac, and the Consejo Nacional de Ciencia y Tecnología (CONACYT) for their support in this endeavour.

My life friends Lucero and Diego, my teammates Rebeca and Edwin, and my former students Victor, Carlos, Tere, Isra, Luis, Camilo, and Irving for never letting me stop asking questions.

Diego and Emma, for always reminding me the importance of telling very long and detailed stories.

Cyntia, for not letting me miss out on the important "chismes" and helping me re-fresh my mind when I was clogged.

Ma, for hanging on the other side of the phone while I read, and re-read, and re-read page after page out loud to you.

Pa, for reading every word, spending long hours giving me your comments, giving me your suggestions, checking in on me to see if I was still working and reminding me to exercise.

Beka, for with the power of your stare and the strength of your paw you reminded me that the book is not all there is in life.

Pedro, for putting up with me for over twenty years now (three dissertations and now a book), making everyday fun, interesting, surprising, challenging, and simply worth living.

My medical and anthropological household made me a medical anthropologist,
my postmodern environment made me an STSer,
my life companion made me a critical thinker,
and my students made me a researcher.
I write this book because of them.

Contents

1

A National Portrait

A constant and loud huuummm.
A huuummm that encapsulates the place allowing you to hear the size of the room.
A never silencing huuummm.
On top of this huuummm, a new sound.
A tap tap tap.
Rhythmically, tap tap tap, tap tap, tap tap tap tap,
a few slightly different tones of "tap".
Each tap a sperm.
Each slightly different tone, a different way to count sperm.
Each tap one sperm passing in front of her eyes,
across a square,
under the microscope,
within the huuummm.
Her eyes moving between squares, catching each sperm, and taping to count it.
Then a bing!
All I could think of was: how on earth can someone count these tiny moving things?
It is like trying to count moving stars!

© The Author(s) 2020
S. P. González-Santos, *A Portrait of Assisted Reproduction in Mexico*,
https://doi.org/10.1007/978-3-030-23041-8_1

Portraits are purposely composed images of people aimed at representing, through the various details they encompass, the social position, psychological characteristics, personality, sometimes even mood of the person being depicted. These images also offer clues about the historical context in which they were created. Hence, given the richness of information they posses, portraits are useful points of departure to explore the cultural aspects of a given historical period. A clear example of this is the portraits painted by the Milanese artist Giuseppe Arcimboldo (1527–1593). He carefully and ingeniously assembled apparently disparate but related elements like plants, flowers, animals, vegetables, and tools of different trades and professions in such a way that they produced a multilayered visual narrative. Looked at from a distance and as a whole, you see the portrait of a person; then, looked at in more detail and with contextual knowledge, we identify visual and rhetorical paradoxes, allegorical meanings, puns, and even jokes. Yet upon closer examination, the narrative of each portrait tells us a story of the scientific, technological, and political characteristics of the period. I took inspiration from his work[1] and wondered: How would the portrait of Mexico's assisted reproduction system look?

Imagine the following elements: a postal stamp circulating in the early 1930s, the symbols of public healthcare institutions and of medical associations, advertisements of fertility clinics displayed in public spaces like highways, buses, and backstreets; newspaper clippings and magazine articles telling the successful stories of people's experience with different assisted reproduction techniques; scientific articles reporting the changes of these techniques; promotional materials like USBs in the shape of a sperm or a pen with a translucent capsule where plastic sperm and ova are swimming. Each of these elements begins to sketch a composite portrait of Mexico's assisted reproduction system.

---

[1]However, his was not the only sort of portrait that inspired this work. I would also like to acknowledge and recommend the work of Chris Jordan (Portraits of global mass culture), Gina Glover (Art in ART), and Helen Chadwick. Their work has been a point of departure, a source of reflection, and interlocutor throughout my years as a professor and a researcher.

Throughout the book, I gather different elements from the field and assemble them in such a way as to paint, through words, a composite portrait of Mexico's system of assisted reproduction, a "repronational portrait" (Franklin and Inhorn, 2016) that depicts the social position and psychological characteristics of the field as well as the political and economic context in which assisted reproduction has been introduced, and has developed and flourished. This portrait highlights the particular social position occupied by the Mexican system of assisted reproduction and the work it does within contemporary culture. It talks about how assisted reproduction has been cutting edge in more ways than the biotechnological, how it has transformed the way Mexicans conceive of and engage with health care, reproduction, the social media, the global market, and even with kinship and family. It exemplifies how assisted reproduction diffused simultaneously among the wealthy population of potential users and within a wider sector of the population who, although perhaps not having the economic means to attain these services, are nonetheless influenced by their presence. This portrait depicts how infertility is no longer presented as a medical condition in need of a cure but as a problematic situation for which the solution is available at these clinics. It is assembled from the different reproductive policies and practices and talks about how this field has been co-constitutive with the process of globalisation and of the establishment of neoliberalism as a political economy. Overall, this portrait shows how assisted reproduction has undergone epistemic and material transformations. The procedures, the practitioners, the users, and the workplaces, as well as the condition itself, have all changed: from being patients to becoming consumers; from addressing only heterosexual couples to catering to same-sex couples and single individuals; from being performed in a medical office to being offered at larger more complex medical facilities; from being done by a single practitioner to requiring a multidisciplinary team of experts. This book offers elements to paint this portrait, it tells the story behind this portrait, it traces these changes, and it sets out the context for what is to come.

Structure

The book is divided into this introduction, followed by two parts with three chapters each. In the remaining of the introduction, I establish the boundaries of the object of this study, the Mexican AR system. Then, I briefly describe the methods and theoretical tools I used to generate and make sense of the information that is at the core of this book and I close this chapter with a preamble to the story of assisted reproduction in Mexico.

Part I, *Origin*, looks at the emergence of the Mexican system of assisted reproduction, specifically at the establishment of its material and epistemic infrastructure. This first part is focused on seeking to understand how Mexico's AR system got to be the size it is today, to work as it does, and to do what it does. It offers a story of medical specialisation and professionalisation as well as an origin story. This first part of the book comprises three chapters. In Chapter 2, *Interest in Sterility*, I look at the people and institutions concerned with infertility and sterility. This chapter covers from the 1940s to the 1960s, and it looks at how sterility became a field of interest for a group of physicians, leading them to create a specialised association and its journal. In *Managing Reproduction*, Chapter 3, I trace the epistemic shift taken by this association, from focusing on sterility to becoming interested in managing reproduction, as part of the family planning campaigns Mexico was engaged in. In Chapter 4, *Turning to Reproductive Technologies*, I look at a second epistemic shift, during which the growing interest in biotechnologies used to assist reproduction flourished in Mexico.

To tell these stories, I go back in time just a few decades. This flashback allows us to see what was happening before the first births or even the first clinics were established, and to see the origins of the system, not from the perspective of the first births but from the perspective of those first engaged in infertility. This chapter asks the following questions: Who was experimenting with ways of curing and overcoming infertility? Who began asking questions regarding assisted reproduction? Why did they become interested in this area? When and how did they begin experimenting with assisted reproductive technologies (ARTs)? And, how did they build the infrastructure to carry out their research in this field?

Part II, *Reproducing Assisted Reproduction*, covers the 25 years after the first successful birth in 1988, period during which the first generation of ART users who had to face the task of making assisted reproduction useable. It focuses on the discursive landscape where ARTs were being discussed and negotiated, in terms of their sociopolitical and moral aspects, with the purpose of understanding how ARTs became socially usable, in addition to biologically and technologically useful. I begin this second part of the book by setting out the conceptual framework, explaining how and why I approach ARTs as technological innovations and as cultural novelties. Then, I offer an overview of the events relevant to the field of assisted reproduction that took place between 1990 and 2010. The first chapter of this part, *The Universe Is Expanding*, looks at how the universe of assisted reproduction expanded from 3 clinics in 1985 to 17 by the year 2000. I highlight two points that are particularly relevant during this period: the configuration of the AR clinic and the establishment of the Latin American registry. In Chapter 6, the *Discursive Landscape*, I argue that through the discourses and narratives offered in different spaces like the media, support groups, trade shows, and recruiting events, assisted reproduction was made socially usable in the sense of making it an acceptable way of forming a family. The book closes with *Contemplating a Repronational Portrait*, a reflexive chapter in which I look at the portrait this book painted from a distance, to appreciate the larger picture and thus see what this book has done, what Mexican AR system has accomplished, and what could we expect to come.

Naming the Mexican Assisted Reproduction System

The Naming of Cats is a difficult matter,
It isn't just one of your holiday games;
You may think at first I'm as mad as a hatter
When I tell you, a cat must have THREE DIFFERENT NAMES.

T. S. Eliot

Indeed, naming is a difficult matter; be it naming a cat, a book, or an object of study. Naming is not a game; it matters. It matters because by conjuring a name to designate the object of our thought, we bring it into being, for us and for those with whom we are in conversation. By naming it, we create it, we delimit and confine it, and we make it manageable. Each name speaks of a specific group of elements set in particular relations (think of your full name and how it indicates your family lineage, the culture you live in, and maybe even your religious heritage). Names evoke different histories and they enable different things. In this book, I name the object of my study: the Mexican Assisted Reproduction System.

As T. S. Elliott illustrates in his poem, the important thing about names is that they can do different things. For example, names can be confusing. In the first written feedback I received of my Ph.D. research project, the reviewer used the abbreviation NRT. I remember staring at those three letters not knowing what they stood for. It took me several minutes to figure out what they meant: New Reproductive Technologies. Ah, yes! I felt terribly silly for not knowing this immediately, particularly since my research project was precisely about reproductive technologies. In 2006, when I was beginning my Ph.D., in vitro fertilisation (IVF)[2] and other techniques used to aid people in reproductive processes were no longer new. Louise Brown, the first child born after being conceived using IVF, was about to turn thirty; and in Mexico, the site for my research, new clinics were opening every year. So, in my eyes these technologies were no longer new. For me, they were simply technologies assisting reproduction or for short, ARTs.

Things change, and so do their names. A decade later, in 2016, the number of infertility clinics offering this sort of assistance had increased, from little over twenty in 2007 to around 100 in 2017. They were targeting other segments of the population beyond the heterosexual couple with a "medical diagnosis for infertility"; many were now catering to single women and men and to same-sex couples. The technologies they offered were also expanding, from artificial insemination (AI), IVF, and

---

[2]Appendix has a list and description of each procedure.

intra-cytoplasmic sperm injection (ICSI), to fertility preservation services and sex selection. Their advertising strategies were becoming more sophisticated. In addition to ads in public spaces like bus stops or printed in magazines and newspapers, they were now using social media outlets like Facebook, YouTube, and Instagram. In these ads, they were presenting information that went beyond describing the different techniques available; they were also talking about financing options, promotions ("20% discount during May" because of mother's day), and guaranteed success programs ("a baby or your money back"). Given the number of clinics and patients in need of specific drugs and materials, specialised pharmaceutical and laboratory supply companies flourished. In the midst of this expanding market, policy makers presented over twenty legislative proposals to regulate these services, none being fully discussed, let alone passed on to a voting stage. As this brief overview suggests, there is a great variety of elements making up the practice of assisted reproduction in Mexico, making the T that stands for technologies in ARTs feel constraining and somewhat deceiving. Assisted reproduction is now larger than the technologies. Drop the T and add an S: and S for System.

Studying Systems

What are systems? How should systems be understood? In order to make sense of the Mexican AR system, I followed the work of Thomas P. Hughes. He had studied the electrification processes in the USA, the UK, and Germany with his initial question being: "how did the small, intercity lighting systems of the 1880s evolve into the regional power systems of the 1920s?" (Hughes, 1986: 283). My question was very similar: How did a group of 31 male physicians studying infertility and arguing against artificial insemination in the 1950s evolve into a system of approximately 100 clinics offering in vitro fertilisation and other biotechnologies to help people have children? Hughes' focus of attention was not so much on how things worked (how a light bulb gives light), but on how things become useful and on how these socio-technical systems evolved (Hughes & Coutard, 1996). Like Hughes, my focus is not on how the different technologies work (IVF or any of the other techniques such as ICSI or AI), but on how these technologies have been worked on, assembled and reassembled, transformed (Callon, 1987: 96) to make them socially *usable and useful* in Mexico. Thinking about a system means asking about its builders, those actors who "nurtured the technical devices by raising funds, circulating publicity, responding to legal requirements, and founding organisations such as manufacturing facilities and utilities" (Hughes & Coutard, 1996: 45), and also about the managers, natural resources, geography, legal aspects, and unexpected events. When all these are articulated, they will set a particular style to the system, which in turn will set out a specific way of working with other systems, and thus of evolving. These articulations are never fully stable; they are dynamic and responsive to changes. Hughes identified seven sequential yet overlapping phases in which different actors emerge and become more or less predominant, changing the way they are articulated and thus the way the system works and what it does. These phases are: invention, development, innovation, transfer, growth, competition, and consolidation. Who were the Mexican AR system builders? Can these phases help understand the story of the Mexican AR system? Did the Mexican AR system go through these phases?

According to Leigh Star (1999), systems do not grow de novo in the void; they are built from and onto a pre-existing base, which in turn they contribute to build and transform. They are embedded into and inside other structures, social arrangements, and technologies; they are part of a community with conventions of practice. Which was the pre-existing base upon which or from where this AR system was built? Which are the communities that claim power over the AR system and how do they claim and use this power? Leigh Star also asks about the master narratives that are embedded in the system's infrastructures, about the way the *other* is constructed, about the work done by those who are not mentioned, about the power distribution within systems, and about the paradoxes in how systems work. What is the master narrative articulated in this AR system?

This book is particularly influenced by certain ideas and methodological suggestions common to the field of science and technology studies (STS). One, the suggestion to avoid establishing a priori borders and distinctions between realms, between what can have agency (do something) and what cannot, between the social and the technical, the macro and the micro, the national and the international. STS suggests observing and listening to how the actors establish, negotiate, and interact with these borders and being attentive to the changes that might occur: "...paying attention to how the borders [...] were drawn by the actors, rather than assuming that these borders are pre-given and static" (Bijker, Hughes, & Pinch, 2012: xvii). Two, STS calls to unpack even the most common concept and be open to all the possible understandings and uses actors might give to them; for example, the content of concepts such as gender, health, body, individual, power, knowledge, reproduction, infertility, etc. should be questioned and not assumed to mean what we think they mean. Three, the push to offer a new narrative. Promethean histories of science and technology already live in bookshelves, stories that emphasise the heroic inventor, the genius engineers, the mad scientist, the magical alchemist, or the exotic artist have already been heard. My aim is to bring in new voices, places, actions, problems, etc. and offer new fuller and thicker descriptions that connect what, on the surface, might seem as disconnected. Four, to recognise that actors are versatile; they tend to do more than what is expected: "...system builders simultaneously had to engineer technological matters ...economic matters...and political matters" (MacKenzie, 2012: 190). Five, the material semiotics branch of STS suggests we focus on the material, because when "we look at the social, we are also looking at the production of materiality. And when we look at materials, we are witnessing the production of the social" (Law & Mol, 1995: 1). Social life is the result of a series of ordered or patterned "networks of heterogeneous materials [...] networks composed not only of people, but also of machines, animals, texts, money, architectures" (Law, 1992). Our task as researchers then, is to explore these networks in order to see how "they come to be patterned to generate effects like organisations, inequality, and power" (Law, 1992); to ask about how networks, webs, or systems hold together and what produces their durability? (Law, 2009); and about what gives them their particular style (Hughes & Coutard, 1996).

Studying Reproduction

This book is not alone. It was not created alone, and hopefully, it will not be read alone. This book is part of, and seeks to contribute to, a family of research projects (books, chapters, articles, special issues, talks, meetings, courses) that are addressing the problem of reproduction in the twentieth and twenty-first centuries.[3] This interdisciplinary field is interested in exploring and charting the role of reproductive technologies (from contraception to IVF and beyond) in the construction of individuals, families, societies, nations, and well, life in general. These projects have offered specific "repronational histories", "national portraits", or "repronational choreographies" (Franklin & Inhorn, 2016), detailed stories of the development and articulation of these technologies in different geographies and following different rationales. These stories suggest that ARTs are not simply technologies that help people have children, but that they can also do other things. For example, they can be political tools to prove how modern and technologically advanced is a region, a country, a health system, or a local medical group (Bharadwaj, 2016; Franklin & Inhorn, 2016; Olavarría, 2018; Tamanini & Tamanini Andrade, 2016; Wahlberg, 2018); they can also be "a technology of social contract" (Mohr & Koch, 2016), a way of responding to the sociocultural values assigned to kinship, gender identity, and lineage.

By focusing on ARTs, these histories have also been able to give detailed accounts of the process of globalisation,[4] of technological diffusion, of global disparities (Gerrits, 2016; Gürtin, 2016; Hörbst, 2016; Simpson, 2016; Whittaker, 2016), of the strength of the neoliberal project (Thompson, 2016), and of the commodification of life, of social

---

[3]These are a few books that have brought together the work of a group of scholars involved in the study of reproduction in many different geographies. This list does not aim at being exhaustive; it is simply inspirational: McNeil, Varcoe, and Yearly (1990), Faye, Ginsburg, and Rapp (1995), Inhorn and Balen (2002), Unnithan-Kumar (2004), Lock and Kaufert (2006), and Straw, Vargas, Viera Cherro, and Taminini (2016).

[4]It is worth reflecting on how reading these national portraits within a global contexts illuminates the biases of how the notion of "the global" is made. Take for example the global rankings of the number of clinics, cycles, or babies born as a result of IVF. These rankings are constructed using data generated either nationally or regionally; however, the logic behind these registries is not the

relations, of the body, etc. This book, for example, tracks the shift from a Malthusian inspired population project, where population policies were presented "in the guise of modernity, development, and linear notions of progress" (Krause & De Zordo, 2012), to the neoliberal project, where the politics of reproduction openly defend and heighten individual rights and well-being (including the rights of the unborn) together with the marketisation of reproduction. Looking at these histories, they show how IVF and reproduction in general are constantly changing in many different ways, in many different directions, and at different velocities. In other words, they show how change is a dynamic, global, yet situated process. These projects have suggested that IVF in particular, but ARTs in general, can be seen as "transformative global technologies" (Franklin, 2013), as a "key symbol of our times [...] an illuminating lens through which to examine contemporary social relations during a very fluid, complex epoch" (Inhorn & Birenbaum-Carmeli, 2008).

It is important to underline that, in general, these projects do not have the purpose of evaluating the success or failure of these technologies in any given context, nor of determining their moral or ethical legitimacy. Instead, they seek to highlight the distinctive characteristics of how ARTs are understood, used at, performed in, and articulated into the local and national cultural contexts. They seek to identify how cultural values, norms, and rationalities are changing as the result of the emergence and dissemination of AR. They are also interested in tracking the process by which new terms, myths, and rituals are created as a result of these articulations (Franklin, 2013; Franklin & Inhorn, 2016; Inhorn & Birenbaum-Carmeli, 2008).

The aim of this book, particularly when put in dialogue with its kin, is to think about the contemporary biopolitics of reproduction and how both the biopolitics and its biotechnologies are making possible

---

same everywhere. For example, the Latin American Registry (RedLARA), in 2018 it evaluated Mexico's performance based on the information given by 32 of the approximately 100 active clinics. Locally, it has been only recently that a governmental body (COFEPRIS) began certifying laboratories (not clinics as a whole). Hence, the numbers used to locate Mexico within a global ranking are very limited; they do not take into account all the clinics, cycles, procedures, or children born as a result of assisted reproduction practices.

a scenario that goes beyond the question of who can or cannot reproduce. Think about Dr. Zhang and his use of spindle nuclear transfer to create a child with three different sources of DNA (González Santos, Stephens, & Dimond, 2018; Zhang, Liu, Luo, & Chavez-Badiola et al., 2016; Zhang, Liu, Luo, & Lu et al., 2017), about Dr. He Jiankui use of CRISPR to fabricate babies that would be immune to HIV (Wahlberg, 2018), and about Dr. Jacques Cohen production of children using ooplasm transfer (Chen, Pascale, Jackson, Szvetecz, & Cohen, 2016; Cohen, Scott, Schimmel, Levron, & Willadsen, 1997). Given that these are all babies born, techniques used, services sought for, the discussion can no longer be (only) about whether these techniques can or should be used, but about what sort of society we *are* manufacturing, how it is going to make sense of itself, what sort of values and ideals is going to guide it, and what future it is creating.

Given all this, this book offers a story of how the Mexican AR system was created and evolved, about the style and patterns of its evolution, and about its material, epistemic, and symbolic infrastructures. It is a book about the messy and complex interactions between different people, artefacts, organisations, and narratives that, by coming together, they enable this particular socio-technical system, a system that seeks to assist people to have the sort of family they want, that is dynamic, and that has a local uniqueness as well as a global similarity. This book is not (only) about an ailment (infertility) or a biotechnology (ARTs) that aids it, nor is it about the experience of patients suffering from infertility and undergoing IVF (although these issues do appear), it is about the way assisted reproduction is made usable in one particular sociocultural context.

Fieldwork

Wednesday 30 January 2007 8:00 am: "Stand there and observe" was the instruction I was given by the head nurse on my first day of fieldwork at a renowned infertility clinic in Mexico City. She had been working at that clinic for many years, and it was clear that she had full control of all the patients, physicians, junior nurses, and embryologists. She knew who, when, how, why, and how many about everyone and everything. I obeyed. I stood there and saw how people went about. Nurses and doctors would pull out folders, flick through them, put them back in; they would come in and out of rooms, they would fold linen, and they would write on charts. Nothing made much sense to me. But as the days passed and I saw the choreography playout, I was able to start making sense of the movements of the dance, to understand their rhythms and tempos. While I stood there and observed, I wondered how and when I was going to be able to slip into the dance so I would not interrupt, so I would not get in the way and cause the full choreography to break down. Standing there and observing, being patient, doing things slowly, holding on to the question as long as I could: this was what worked for me. The way I was to engage with my field had to make sense in the field itself. Reproduction is a dance, even when it is medically assisted. It is all about rhythm and choreography (Thompson, 2005), and so had to be my ethnography. Knowing when and how to step in was crucial. Soon enough I was able to participate, first as a shadow of other dancers (the nurses, the physicians, the med students, the embryologists) and then as a dancer in my own right. I was assigned tasks: "could you talk to this patient while the doctor finishes with this other one", "could you help the nurse explain the procedure to the patient", "could you accompany her husband while she wakes up".

Studying ARTs has meant talking to healthcare professionals, working with lawyers, interviewing equipment distributors, shadowing embryologists, meeting marketing experts, and policy makers. It has meant spending time at clinics, sitting at medical conferences, observing patient groups, touring distribution hubs, learning about incubators and culture mediums, listening to air filters and laboratory tally counters, reading journals, trying to understand the process of policy making, watching and analysing tv shows, movies, advertisements, websites, and Facebook pages. Throughout these past ten years, I have gone from an observer to a participant observer, to an ethnographer (de la Cadena, Lien, Blaser, Jensen et al., 2015). As an ethnographer, you engage and pay attention with all your senses. You generate and collect notes, objects, sound bites, smells, and feelings; you establish relationships. You observe everyday activities and engage in conversations about all sorts of topics, many of which could be considered out of the realm of the question or topic under research, but which are crucial for making connections and for generating questions. It is these multiple and multilayered interactions that distinguish ethnography from other sorts of qualitative engagements. Ethnography, for me, is an obsession with finding the thinnest of threads and seeing how they are woven together to make up fabrics with beautiful intricate patterns; it is wanting to listen to all the stories that surround these threads and knots; it is connecting and caring about these threads, knots, and the beings and places that make them. Ethnography means caring, and caring means being responsible (Haraway, 2008), caring for your informants, for their stories and perspectives, assuming responsibility for the question you ask, for the stories you tell, and for how you affect the field of which you are now a part. Ethnography also means establishing a long and engaging relationship with the people in your field. This long engagement means that in the course of ten years, the field and I as a researcher have changed, as have the questions that guide my work.

Between 2007, when I began my fieldwork, and 2019, when I finished writing this book, the number of clinics increased, their use of advertising strategies diversified, the tools and techniques used in the embryo lab became more complex, and the availability of certain services changed (e.g. surrogacy). To illustrate this, in 2007 only some clinics had websites, but in 2019 very few do not have a website, and many also have other social media accounts such as Facebook, YouTube channels, Instagram, and Twitter. Likewise, between 2007 and 2010 I noticed that patients used chat rooms to share their experiences; in 2017 this sort of sharing was done through Facebook, Instagram, and YouTube.

During these years, my engagement with the field also changed. When I began, I knew no one; my understanding of the science was limited; I had never been to an AR clinic, much less in an operating room; and there were very few studies to use as a starting point. Those that existed either focused on the legal conundrums or addressed only the psychological aspects of infertility and the use of ARTs. Furthermore, due to the peculiarities of Mexico, there was very limited statistical data about the number of clinics, physicians, procedures, or babies born after an AR cycle. Thus, my first task was mapping the field of assisted reproduction and identifying its actors (González-Santos, 2011). I began by creating a database of the active clinics; then, I focused on the dynamics within the clinics; I explored the various efforts to establish a regulatory body to norm these services; and, following the suggestion of the nurses and patients, I turned to see how infertility and ARTs were represented in the media. Upon concluding this first stage of my research, I decided to delve deeper into the emergence and development of the field's infrastructure[5]; that is what this book is about.

* * *

The stories I tell in this book are about the people, the institutions, the ideas, the policies, and the knowledge that, when assembled, make up the Mexican AR system. But these stories are also about Mexico, about its government, its inhabitants, and their stories. Studying and thinking about the AR system has been a way of learning about my country, about the materials, practices, and connections that make it and allow it to be. Now, I want to present these to you. Having this dual intention—presenting both the AR system and the country—as well as having an international audience in mind,[6] has guided the decision process

---

[5]My interest in the system's infrastructure was sparked by several ethnographic moments. For example, when I heard a team of AR specialists wondering if the low success rates they were having during that particular cycle could be due to problems with the handling of the culture medium at the customs agency. Then, when at a conference, I listened to physicians talking about how the differences in healthcare systems, cultures, and medical practices between the USA, Europe, and Mexico, explained the different ways in which assisted reproduction was practised in each context, and then again when I saw patients sharing experiences and suggestions about how and where to get hold of certain drugs that were difficult to access. My interest in the historical development of the community came from wanting to explore how the founders of the medical association that today gathers the physicians practising assisted reproduction, who were actually against artificial insemination, changed their minds to accept all sorts of assisted reproductive technologies (ARTs)?

[6]Having a non-Mexican and non-Spanish speaking audience in mind meant I had to make particular decisions in terms of the narrative and structure of this book. How often did I have to remind the reader that what I was saying pertained to Mexico? Having to do so was something I felt was necessary very often, but then wondered how often other authors do this, particularly when their research site is the USA. Likewise, where should the original (i.e. in Spanish) name of a book, journal, association, institution, organisation, or governmental body go and where the translation (into English)? Should I say, for example, Secretaría de Salud (Ministry of Health) or the other way around, Ministry of Health (Secretaría de Salud)? Should I carry out the entire narrative doing this or just introduce the name in Spanish once and then use the

of what I put in and what I leave out. Some may think that the stories I tell have "too much detail", but I feel the need to offer these dense descriptions that sometimes take me far away in time, because through them I am trying to show how this AR system is part of and is located within Mexico, and how Mexico is part of and is located within the AR system. Hopefully, I will be able to take you far away in time and place and safely bringing you back to the object of this book: Mexico's AR system. The first trip in time starts here, in post-revolutionary Mexico. I hope you enjoy it.

Puericulture: Managing and Assisting Reproduction

> He asked me when I had begun, when my life had begun. I answered, 'when my mother decided she wanted a child, a girl, and she would call her Sandra'. He said 'No', 'you never begun, life is simply transformation'. (First Interview, Field Notes, 2007)

Stories do not have a true beginning either; we choose when to begin them. I have chosen the post-revolutionary period because it is then that the conditions of possibility for this AR system were laid out—specifically the idea that reproduction can be managed by individuals through science, and that the state can and should have a say in people's

---

English version? What about translations of quotes, should I offer the original version as well as the translation? Does any of this matter? I believe it does matter, both in political and aesthetic terms, but I am not sure which is the answer or solution to these questions or problems. I decided, not entirely convinced though, that I was going to place the reader as the priority, thus make the narrative and structure as friendly as possible. So, this is how the book is written: I introduce the name in Spanish first (and offer the English translation in parenthesis, with the Spanish acronym when given the case) but then use the English name or Spanish acronym throughout the book. I have translated all quotes that were originally in Spanish myself and only offer the original quotes upon request or downloadable from the electronic version. As I said, I am not entirely sure if this is the best option this is why I wanted to offer this reflection as a way to contribute to the discussion regarding the politics of knowledge production (see, e.g., Pérez-Bustos, 2017; Rodriguez Medina, 2019).

reproductive practices. Historians consider the post-revolutionary period to cover, roughly, between the 1920s and the 1930s. During this period, the country was governed by a series of presidents, each ruling for no more than a few years: Alvaro Obregón (1920–1924), Plutarco Elías Calles (1924–1928), Emilio Portes Gil (1928–1930), Pascual Ortiz Rubio (1930–1932), Abelardo L Rodriguez (1932–1934), and Lázaro Cárdenas del Rio (1934–1940). It was president Cárdenas who first fulfilled the six-year term, a *sexenio*, initiating with this a new period in Mexico's history known as the stabilisation and development period.

The Mexican Revolution, fought during the 1910s, left the country with an eroded population and with a devastated social, economic, political, and material infrastructure. Hence, the main goal of the post-revolutionary period was to rebuild and redesign Mexico, to transform it into a modern country. Key to this modernising project was, first, to reduce the sociopolitical power of the Church and to increase the sociopolitical power of the State. The second goal was to move from an agrarian to a manufacturing economy and thus achieve a higher level of industrialisation. The third goal was to repopulate the country with a healthy population. The very variable and inaccurate approximations suggest that during the revolution the country lost between 1 million and 2.5 million people (between 5–7% of its population), due to violence, high migration, and the epidemics of typhoid fever, yellow fever, and smallpox (Alba-Hernandez, 1976; Agostini, 2007; Mateos Fournier, 1964; Guerrero, 1954; Gutiérrez-Sánchez, 2000; Mendoza García & Tapia Colocia, 2010; Mier y Terán, 1991). Infant and maternal death were high, and they were associated with prenatal issues, such as mothers with sexually transmitted diseases (STDs) and their abuse of toxic substances like alcohol, morphine, and a combination of morphine and scopolamine used to relieve childbirth pains,[7] and to perinatal factors, such as complications at birth and poor conditions during the first few years of the child's life. The medical community considered these pre-, peri-, and post-natal deaths to be related to poverty and ignorance,

---

[7]Known as Twilight sleep.

and located the responsibility for these within the mother–child dyad (Alanís, 2015; Espinosa de los Reyes-Sánchez, 2016). The father and the rest of the family were not mentioned as equally important actors in the child's conception, the conditions of its birth, or upbringing.

In this context, a hygienic project was considered fundamental for the country's reconstruction. In the realm of reproduction, many of these projects focused on the pregnant woman, the newborn, and the child.[8] These laws established who should reproduce, how they should reproduce, who should manage reproduction, where it should be managed, and what knowledge set should guide its management. Take, for example, the Ley de Relaciones Familiares (Family Law), which is part of the 1917 Constitution. This law reframed marriage as a voluntary civil contract, established by the State (and not the Church), that could be rescinded, and with three very clear purposes: the mutual support of the married couple, bearing healthy children, and providing them with a healthy environment and an education.[9] To attain these last two goals, the Family Law stipulated that in order to attain a marriage license, the couple had to present a medical certificate to the judge, confirming that they had been checked by a health professional and that they were healthy, free from the "mortal trinity" (tuberculosis, syphilis, and alcoholism), impotence, madness, or any other chronic or incurable disease. The argument justifying this requirement was that if the couple had any

---

[8]This contrasts with the projects held by the Church and the civil organisations which were doing charity work prior to the Revolution. Those had mostly focused on the homeless, the mentally ill, and the orphans. However, these ideals also fed other sorts of policies and projects put forth during this period, some of which received money from abroad. For example, President Álvaro Obregón (1920–1924) implemented a series of campaigns to improve the sanitation and hygiene of the major cities and ports, some of these were sponsored by the Rockefeller Foundation.

[9]This law, written by the Congreso Constituyente (the group of people who wrote the constitution lead by President Venustiano Carranza), sought to disentangle marriage from religion and, by establishing the possibility of divorce, they hoped that more people would marry. This in turn would help reduce the number of illegitimate children and it would help protect women from abusive situations, since it granted them more rights (e.g. the husband is no longer the wife's legal guardian and representative although he still needs to grant her permission to work, study, or establish her own business) (Adame Goddard, 2004). It is worth noting that Benito Juarez had already passed the Ley de Matrimonio Civil (the Civil Marriage Act) in 1859, which stipulated that marriage was a civil contract making religious marriages not legal.

of these conditions, the marriage would not be able to produce healthy offspring, which in the end was the purpose of marriage (Stern, 2002). Furthermore, they were also asked to attend premarital talks during which they received instructions concerning basic hygiene practices to care for the newborn.

This regime of reproduction had a pronatalist agenda (Gutmann, 2009; Mejia Modesto, 2007; Stern, 2002; Vallarta Vázquez, 2005; Zavala de Cosío, 1992), with an eugenic perspective, and it used puericulture as the method to implement them (Stern, 2002). Puericulture was the "science with the objective of researching and studying the causes relative to the conservation and improvement of the human species" (Espinosa de los Reyes, 1921, in Espinosa de los Reyes-Sánchez, 2016). It must be understood within the context of the Mexican eugenics perspective. This perspective had very specific premises. First, it was a pronatalist project that, while focused on the individual, had as its overall purpose building a healthy nation. Second, it was a perspective which considered that certain types of undesirable traits could be avoided or modified if the right conditions were met (Saade Granados, 2004). These conditions were usually related to better education, urban infrastructure, hygiene habits, and adequate public policies. Third, it emphasised the role of the state and of philanthropic agencies as the ones responsible for ensuring that the necessary conditions were met to avoid passing down these negative traits. Fourth, it viewed certain segments of the population as illiterate, ignorant, insalubrious, and suspicious of physicians, medicine, and the government institutions. Fifth, it re-signified mothers as powerful, even over their husbands and fathers, if (and only if) they carried out their (hygienic and eugenic) motherly duties. For example, physicians such as Alfredo M. Saavedra, author of the book "Eugenesia y Medicina Social", (Eugenics and Social Medicine, 1934) questioned the traditional patriarchal authority granted to the father–husband figure, framing them as lazy, criminal, and brutal; and sought to substitute the father–husband figure with other male figures such as the physician, the state, and even the school (Saade Granados, 2004). These were depicted as trustworthy and as having the woman, the child, and the nation's interest in mind.

The sum of these premises rendered reproduction, maternity, and child rearing as matters that could be medically and scientifically managed. To be able to follow this "scientific maternity", as it was commonly referred to by professionals, women needed constant medical advice and supervision (Agostini, 2007; Saade Granados, 2004; Stern, 2002). The state trusted eugenicists, puericulturists, and hygienists with the task of reorienting the behaviour of mothers. One specific task was to eliminate the midwives, who were seen as "the remnant of an unhealthy and superstitious past", and instead promote the use of gynaecologists and nurses. In sum, puericulture became key in the solidification of gynaecology and paediatrics in México, for the establishment of the "Gran Familia Mexicana" with the mother–child dyad at its core, and for the establishment of a discourse that assigned responsibility to both the mother and the state in protecting Mexican children and creating a healthy population. Through science, technology, and infrastructure, the state was trying to remodel and modernise the country. For this to happen, science itself, and medicine as one instance of it, had to also be modernised (the project of modernising medicine will be discussed in the following chapter).

A key figure in this period was the obstetrician Isidro Espinosa de los Reyes. Among other things, he promoted a legislative package that would help improve the conditions of women prior, during, and after pregnancy. For example, he proposed giving working pregnant women two-month maternity leave, one before and one after birth, to properly prepare and care for the newborn baby. He encouraged the establishment of day care centres and half boarding schools, so children would be looked after while parents worked. He suggested regulating the accessibility to toxic substances such as alcohol, tobacco, and morphine in order to reduce the intake of these elements prior and during pregnancy. He fought for the development of urban infrastructure such as pavement and sewerage. And he created a school for training nurses, so they could supervise the living conditions and habits of pregnant women and newborn children to make sure they complied with the eugenic and scientific ideas. Particularly pertinent to this story was his ambitious project to establish Centros de Higiene Infantil (Child Hygiene Centres), which he built in several neighbourhoods of Mexico City: Peralvillo, Santa María la Ribera, Balbuena, Doctores, Centro, and Las Lomas. The task of these centres would be to offer pre-, peri-, and post-natal services to women and children in need. As a way to prepare those who would carry out these services, he created the Escuela de Puericultura. The overarching message of the Hygiene Centres and the public policies that accompanied them was the promotion of a conscious maternity (maternidad consciente) and of puericulture practices in those who were reproducing (Alanís, 2015; Espinosa de los Reyes-Sánchez, 2016; Stern, 2002: 303).[10] This meant that they encouraged women who were considered unfit to reproduce (because they had an undesirable trait, either biological or social) to voluntarily avoid reproducing, or in some cases they were given forced sterilisation.

One of these Hygiene Centres was located at the corner of Montes Urales and Prado Sur, in the affluent borough called Las Lomas de Chapultepec (Fajardo Ortiz & Sánchez González, 2005). Inaugurated on 27 June 1929, the name of this healthcare facility would be: Casa de Maternidad de las Lomas de Chapultepec[11] (Lomas de Chapultepec Maternity Home). This particular hygiene centre received its funding from three sources: the Secretaria de Salud (Health Ministry), private donations, and from a special tax placed on the local postage stamp

---

[10]One way to promote this "conscious maternity" was through a book specifically targeted to women, called Doña Eugenesia y Otros Personajes. It was a compilation of stories, short dialogues, and poems, created by students (physicians and nurses) of the School of Health and Hygiene, based on their fieldwork experience, and edited by their professor Dr. Manuel González Riviera. Among the conditions addressed in this compilation were: gonorrhea, tuberculosis, typhoid fever, and malaria (O'Hara, 1944). This book can be considered as a precursor of the telenovelas (soap operas) which, as I will detail in further chapters, were used during the family planning campaigns in the 1970s and 1980s, and then in the 2000s as a way of making assisted reproduction usable.

[11]This Hygiene Centre was established where the Casa de Salud del Periodista was located, a charity healthcare facility founded, in 1921, by Félix Flugencio Palavicini, a topographic engineer, a professor, a journalist, and a politician. Flugencio Palavicini participated in the writing of the 1917 Constitution (rumours say he promoted the divorce law since he himself wanted to divorce), he supported women's right to vote, he founded two newspapers (El Precursor and El Universal), and, as an active eugenicists with a particular interest in puericulture, he organised, financed, and promoted, the First National Conference on Children (2–9 January 1921). At this event, Isidro Espinosa de los Reyes presented his paper "Apuntes sobre puericultura intrauterina" (Notes on in-uterus puericulture). In 1952, after Espinosa de los Reyes had passed away, this Hygiene Centre was renamed to honour its founder, becoming the Casa de Maternidad de las Lomas de Chapultepec Isidro Espinosa de los Reyes.

which read: "Proteja a la infancia. Haga Patria" ("Protect children, make nation") (Espinosa de los Reyes-Sánchez, 2016). When President Luis Echeverria (1970–1976) took office, he expropriated the plot of land where the Casa de Maternidad de las Lomas de Chapultepec Isidro Espinosa de los Reyes was located, tore it down, and in its place built a new hospital under the directorship of Eduardo Jurado García (Morales Suárez, 2010, 1999). The name of this new hospital was the Hospital de Perinatología (Perinatology Hospital). The act of renaming the hospital reflected a change of times. In the 1970s, Mexico was a very different country from the one in which the Casa de Maternidad had been created. The healthcare landscape in the 1970s was populated by private hospitals (Hospital Español, 1932; American British Cowdry Hospital, 1952), a series of social security schemes (IMSS, 1943; PEMEX, 1943; ISSSTE, 1960), as well as state-run specialised hospitals (Children's Hospital, 1943; Cardiology Hospital, 1944; Cancer Hospital, 1946; Nutrition Hospital, 1946; Pulmonary and Respiratory Disease hospital, 1959), one of which was Perinatología (Frenk, 2002). These state-run hospitals had four main purposes: research, teaching, advising the government, and serving the community by offering highly specialised attention for those who did not have access to the social security schemes. A decade later, in 1983, President Miguel de la Madrid turned it into an independent decentralised institution. The new Instituto Nacional de Perinatología (National Perinatology Institute, INPer also referred to as Perinato) would be directed by Samuel Karchmer (Frenk, 2002). Decades later they renamed this Institute to, again, remember and honour its long history as a healthcare facility. Thus what once was the Casa de Salud del Periodista (see footnote 10), now is the Instituto Nacional de Perinatología Isidro Espinosa de los Reyes (INPer) (Ahued, 2004; Espinosa de los Reyes, 2008).

Although from the beginning the healthcare facility located at the corner of Montes Urales and Prado Norte was framed as a facility that should be catering to the disenfranchised, this area, Las Lomas de Chapultepec, has always been an affluent neighbourhood. The area houses the richest of the Mexican elite; it has limited public transport access (only by bus) and very few parking spaces. Consequently, those who have to use public transportation to go to the INPer face logistical problems. They can either take a taxi, which can be costly, or a bus from the nearest metro station and walk. Recently, in 2016, the area that connects the metro station and the hospital has been re-urbanised, making the access much more suitable and pedestrian-friendly, and less dangerous, than it used to be. However, when I was conducting the first part of my fieldwork between 2006 and 2010, reaching this hospital using public transport was a serious ordeal.

This place, this corner, has persisted as a healthcare facility since 1921. It has lived through changes of names, of bricks, even the change of ownership. It has endured the transformation from being a charity house, catering to a profession (Casa de Salud del Periodista); to being a welfare facility, looking after mothers and children in need (Hygiene Centres Casa de Maternidad de las Lomas); to becoming a healthcare institution (Hospital de Perinatología); to reaching the level of first-class national biomedical institution (INPer). It has gone from following a welfare model, where the state and the rich are responsible for helping to produce a healthy nation, to a neoliberal system, where those who have less tend to pay the most; from attending all those in need, to those in need with tertiary-care conditions only; from openly following an eugenics views to openly biomedical. It has participated in establishing and re-establishing regimes of reproduction by stating who is eligible to receive the health care they offer and the characteristics of this service. Amidst these changes, some ideas have remained. The different healthcare facilities that have been located at the corner of Montes Urales and Prado Sur have all had the same population in mind: the woman and the child. The exclusion of the father figure has not been an uncommon practice. As examples, I want to reflect on three images used in three cases discussed here: the postal stamp, the symbol of the INPer, and the symbol of IMSS.

The Mexican Healthcare System

Although not explicitly, this book is also about the Mexican healthcare system, about how it was established, fragmented, developed, strengthen, and weakened in less than 100 years. The system of assisted reproduction is an example of this development. Due to this (and to the fact that this book, being in English, is directed to an international readership), I consider it important and useful to offer an overview of the Mexican healthcare system. When reading this, keep in mind that, with every sexenio there are changes to this system, so what I write today will probably have certain changes by the time this book has gone to press or reached your desk. Having said that, what most probably will not have changed is its fragmented state, its amazing capacity to juggle the amount of people it cares for in the most uneven and unexpected situations. The system is fragmented and imbedded with political struggles that have little to do with health, nevertheless children are born in it, sick people are cured by it, injured are rehabilitated, and many people from abroad come to seek their services.

Mexico's health system is the result of a hundred years of infrastructure building, a period that spans from the post-revolution period (1920–1940), the stabilisation and development period (1940–1982), to the neoliberal period (1982 to date). Throughout these years, a series of independent hospitals (both private and public), individual physicians and politicians, and unions of workers emerged and gathered to create a complex, dynamic, flexible series of healthcare sub-systems. While technically each sub-system offers its own set of services, has its own eligible user, its own financing scheme, its own administrative structure, and its own history, they sometimes overlap in terms of the medical staff (there are physicians who frequently work in two sub-systems) and the users (patients frequently use more than one system, e.g. their work-related social security system plus a private healthcare insurance or a private hospital).

For descriptive purposes, I will begin by dividing these systems into two branches: the social security system and the healthcare services. The first difference between these is in terms of what they offer. As their names suggest, the social security system offers more than just healthcare services; it also offers (or offered) retirement pensions, maternity leave payments, accident and life insurance, some also include housing loans, cultural spaces, children's day care facilities, and more (although in recent years some of these elements have been reduced or are under threat of being cut out altogether). The second difference is eligibility, which is determined by employment. People employed in the formal sector (industry and government) are eligible for the social security system, while the rest of the population, i.e. people who are self-employed, who work in small businesses, in the informal sector, farmers, and the unemployed, can only access healthcare services, which can be either public or private (Santacruz-Varela, 2010). The third difference has to do with accessibility. Overall, the Mexican health system is unevenly distributed across the country. While some states, and specifically some cities, have a wide range of services, others are left practically without access.

Today, AR services are offered in all three healthcare systems (private, work related, and public or government related), although the way they are offered varies a great deal between each system due to differences in their operations and physical layouts. However, the majority of AR clinics and services are part of the private healthcare system. Private AR services have the fewest restrictions in terms of patient eligibility (they offer services to heterosexual and homosexual couples, single women, and women over 45), the broadest spectrum of protocols (from homologous artificial insemination to surrogacy including gamete and embryo donation), the widest spectrum of coverage (there are clinics in over 17 states), the highest prices, and many financing schemes to pay for them. The public and work-related services, however, cover a much broader socioeconomic spectrum of the population, yet have more restrictions in terms of patient eligibility and regarding the types of protocols they offer (usually, they do not offer surrogacy), and they have a reduced number of clinics. The public, private, and work-related AR services as a whole seem to cover a wide spectrum of society, although some sectors might be underrepresented. The variety of patients I encountered in each site indicates that, although to different degrees, assisted reproduction is being inserted into various social worlds simultaneously.

Details of a National Portrait

This book opened with a brief description of an imagined portrait of Mexico's system or assisted reproduction. I want to go back to this image and focus on some of the elements of this portrait to see what specific stories they tell. For now, let's focus on the postal stamp, the symbol of the INPer, and the icon of the IMSS.

The Postal Stamp. The one cent postal stamp used to gather funds for the services of the Casa de Maternidad de las Lomas, circulated between 1929 and 1935. It was a monochromatic stamp depicting the profile of a woman looking down at her child, cuddled in her arms while looking away. The legend says "Correos Mexicanos. 1c Proteja a la Infancia. Haga Patria" which translates as "Mexican Post. 1cent. Protect infancy. Be a patriot". Who is protecting infancy? How is being a patriot? One way of reading the visual, textual, and relational message is that the mother, the (male) State who manages the Hygiene Centres, and the wealthy Mexican who is the buyer of the postal stamp, since this stamp was issued explicitly for the area of Las Lomas, are the ones active in being patriotic by protecting the child.

INPer's Symbol. The icon of the Instituto Nacional de Perinatología (INPer) is a Mixtec image chosen by Eduardo Jurado García from the Zouche-Nuttall codex (dated between the fourteenth and fifteenth centuries, and linked to the Mixtecos, an indigenous group from the area of Oaxaca) (Garza Gordoa, 2012; Morales Suárez, 1999; Villa Roiz, 1997). This particular ideogram shows the princess Tres Pedernal Serpiente Emplumada giving birth to her daughter in the traditional kneeling posture of birthing, the umbilical cord is still attached linking mother and daughter, under them is a purple figure which some interpret as the placenta (Morales Suárez, 1999) and on the Princess' head, the feathered serpent. Jurado García chose this image for two reasons: the position the woman is giving birth in and the serpent. Jurado García describes the Princess' position as a position that facilitates the birth process and that is the traditional posture for birthing among Mixtec (as well as many other cultures) (Morales Suárez, 1999). However, it is definitely not the position in which women (are allowed to) give birth daily at this particular biomedical institute, regardless of their culture and beliefs. It is interesting to question then, why use this particular image as the identity image of this institution, an image that depicts the traditions this institution is there to change? The second element is the circular motive constructed by the serpent. Jurado clarifies that he uses a free interpretation to incorporate the serpent into the equation of justification. Although traditionally within the study of

Mesoamerican cultures the serpent has been associated with sexuality, fertility, and water, he draws on the Galean relationship between the serpent and medicine and says that it can thus be understood as "The medical action of protecting the birthing as to avoid any harm to the new born baby and its mother" (Morales Suárez, 1999). To this, he adds that the circle which envelops the mother and child is taken in the INPer's icon to represent "everything that surrounds the birth of a new being", statement that can define what perinatology is and thus the purpose of this Institute (Garza-Gordoa, 2012) as stated by its forefather, Isidro Espinosa de los Reyes.

The IMSS Icon. Framed as one more effort to fulfil the promises set out by the Mexican Revolution, in December 1942 President Manuel Ávila Camacho presented before the Congress a new law that would establish the mechanisms to look after and protect the needs of the working class: "their existence, their salary, their productive capacity, and the tranquility of their family". This law was swiftly approved, and on 19 January 1943, the Ley del Seguro Social (Social Security Law) was officially published. The Instituto Mexicano del Seguro Social (IMSS) emerged from this law. A year after its foundation, its director, Ignacio García Téllez, hired Salvador Zapata Argüello to design the Institute's logo. The main elements of this logo are an oversized eagle swaddling a mother who is cradling and breastfeeding her child. In 1945, the slogan "Seguridad para todos" (Security for all) was added. The image is not innocent. We must remember that the foundational myth of Tenochtitlán, the Mexica-Aztec metropolis located in the valley of Mexico, has the eagle as a centrepiece.[12]

---

[12]Due to problems with one of their gods, Malinalxochitl, the Mexica-Aztecs people had to leave their original land and head south. Some of their other gods assured them they would find a fertile land where they would be able to build a great city. For years, they travelled in search of such a place, until they arrived to a beautiful valley surrounded by mountains and two volcanos, a lake dotted with islands, and an extremely fertile land which gave up to four crops a year. The sign they had been waiting for made its appearance: a majestic eagle sitting on top a tenochtli (a green plant with red flowers and a sweet watery fruit, the Nopal), eating a serpent. That would be their new home; that would be the great city of Tenochtitlán.

These images and myths, although from different times and people, have shaped central Mexico's culture. The overwhelming power of the State, represented sometimes by the eagle (IMSS) and other times by the legend (Postal Stamp and INPer), is strengthened by the absence of the father in all these images. It is not the father who will take care of the family—of the mother and her child—it is the male state. Also important in these visual narratives is the place assigned to science. As mentioned in the beginning of this section, Mexico's modernising project held science as one of its pillars.

References

Adame Goddard, J. (2004). EL régimen revolucionario del matrimonio civil. In *EL matrimonio civil en MÉXICO* (1859–2000) (pp. 35–82). Instituto de Investigaciones Jurídicas—UNAM. Retrieved from https://archivos.juridicas.unam.mx/www/bjv/libros/3/1362/4.pdf.

Agostini, C. (2007). Las mensajeras de la salud. Enfermeras visitadoras en la Ciudad de MÉXICO durante la década de los 1920. *Estudios de Historia Moderna y Contemporánea de México, 33*.

Ahued Ahued, R. J. (2004). Semblanza del Doctor Isidro Espinosa de los reyes. *Perinatología y Reproducción Humana, 18*(4), 205–207.

Alanís, M. (2015). Más que curar, prevenir: surgimiento y primera etapa de los Centros de Higiene Infantil en la Ciudad de MÉXICO, 1922–1932. *História, Ciências, Saúde-Manguinhos, 22*(2), 391–409. https://doi.org/10.1590/S0104-59702015005000004.

Alba-Hernandez, F. (Ed.). (1976). *La población de MÉXICO*. México: Centro de Estudios Económicos y Demográficos, El Colegio de México.

Bharadwaj, A. (2016). *Conceptions: Infertility and procreative modernity in India*. New York: Berghahn Books.

Bijker, W. E., Hughes, T. P., & Pinch, T. (2012). *The social construction of technological systems: New directions in the sociology and history of technology*. Cambridge, MA: MIT Press.

Callon, M. (1987). Society in the making: the study of technology as a tool for sociological analysis. In W. E. Bijker, T. P. Hughes, & T. Pinch (Eds.), *The social construction of technological systems: New directions in the sociology and history of technology*. Cambridge, MA: MIT Press. Retrieved from http://search.ebscohost.com/login.aspx?direct=true&scope=site&db=nlebk&db=nlabk&AN=457446.

Chen, S. H., Pascale, C., Jackson, M., Szvetecz, M. A., & Cohen, J. (2016). A limited survey-based uncontrolled follow-up study of children born after ooplasmic transplantation in a single centre. *Reproductive BioMedicine Online, 33*(6), 737–744. https://doi.org/10.1016/j.rbmo.2016.10.003.

Cohen, J., Scott, R., Schimmel, T., Levron, J., & Willadsen, S. (1997). Birth of infant after transfer of anucleate donor oocyte cytoplasm into recipient eggs. *The Lancet, 350*(9072), 186–187. https://doi.org/10.1016/S0140-6736(05)62353-7.

de la Cadena, M., Lien, M. E., Blaser, M., Jensen, C. B., Lea, T., Morita, A., ... Wiener, M. (2015). Anthropology and STS: Generative interfaces, multiple locations. *HAU: Journal of Ethnographic Theory, 5*(1), 437–475. https://doi.org/10.14318/hau5.1.020.

Espinosa de los Reyes Sánchez, V. M. (2008). Datos biográficos del Dr. Isidro Espinosa de los Reyes. *Boletín Mexicano de Historia y Filosofía de La Medicina, 11*(2), 64–67.

Espinosa de los Reyes Sánchez, V. M. (2016). La asistencia materno-infantil en México entre 1921 y 1930 por parte del Departamento de Salubridad Pública. *Gaceta Médica de México, 152,* 231–245.

Fajardo Ortiz, G., & Sánchez González, J. M. (2005). La medicina mexicana de 1901 a 2003. *Revista Latinoamericana de Patología Clínica y Medicina de Laboratorio, 52*(2), 118–123.

Franklin, S. (2013). *Biological relatives: IVF, stem cells, and the future of kinship.* Durham: Duke University Press.

Franklin, S., & Inhorn, M. C. (2016). Introduction. *Reproductive Biomedicine & Society, 2,* 1–7. https://doi.org/10.1016/j.rbms.2016.05.001.

Frenk, J. (2002). *Los institutos Nacionales de Salud de Mexico.* México: Secretaria de Salud.

Garza-Gordoa, M. (2012). Imagen visual del símbolo del Instituto Nacional de Perinatología Isidro Espinosa de los Reyes. *Perinatología y Reproducción Humana, 26*(2), 133–137.

Gerrits, T. (2016). Assisted reproductive technologies in Ghana: Transnational undertakings, local practices and 'more affordable' IVF. *Reproductive Biomedicine & Society Online, 2,* 32–38. https://doi.org/10.1016/j.rbms.2016.05.002.

Ginsburg, F. D., & Rapp, R. (Eds.). (1995). *Conceiving the new world order: The global politics of reproduction* (p. xii). Berkeley: University of California Press (cloth and paper).

González-Santos, S. P. (2011). Space, structure and social dynamics within the clinical setting: Two case studies of assisted reproduction in

Mexico city. *Health & Place, 17*(1), 166–174. https://doi.org/10.1016/j. healthplace.2010.09.013.

González Santos, S. P., Stephens, N., & Dimond, R. (2018). Narrating the first "three-parent baby": The initial press reactions from the United Kingdom, the United States, and Mexico. *Science Communication, 40*(4), 419–441. https://doi.org/10.1177/1075547018772312.

Guerrero, D. C., (1954). Aspectos Sociales de la Esterilidad. *Estudios Sobre Esterilidad, 5*(1).

Gürtin, Z. B. (2016). Patriarchal pronatalism: Islam, secularism and the conjugal confines of Turkey's IVF boom. *Reproductive BioMedicine and Society Online, 2*, 39–46. https://doi.org/10.1016/j.rbms.2016.04.005.

Gutiérrez-Sánchez, S. (2000). *Transición de la alta a la baja fecundidad en México.* Cuadernos de investigación, Cuarta Época, No. 12. Universidad Autónoma del Estado de México, México.

Gutmann, M. (2009). Planning men out of family planning: A case study. *Sexualidad, Salud y Sociedad—Revista Latinoamericana*, No. 1, pp. 104–124.

Haraway, D. J. (2008). *When species meet.* Minneapolis: University of Minnesota Press.

Hörbst, V. (2016). 'You cannot do IVF in Africa as in Europe': The making of IVF in Mali and Uganda. *Reproductive Biomedicine & Society Online, 2*, 108–115. https://doi.org/10.1016/j.rbms.2016.07.003.

Hughes, T. P. (1986). The seamless web: Technology, science, etcetera, etcetera. *Social Studies of Science, 16*(2), 281–292.

Hughes, T., & Coutard, O. (1996). Fifteen years of social and historical research on large technical systems. An interview with Thomas Hughes. *FLUX Cahiers scientifiques internationaux Réseaux et Territoires, 12*(25), 44–47.

Inhorn, M. C., & Balen, F. (2002). *Infertility around the globe: New thinking on childlessness, gender and reproductive technologies.* Berkeley: University of California Press.

Inhorn, M. C., & Birenbaum-Carmeli, D. (2008). Assisted reproductive technologies and culture change. *Annual Review of Anthropology, 37*(1), 177–196. https://doi.org/10.1146/annurev.anthro.37.081407.085230.

Krause, E. L., & De Zordo, S. (2012). Introduction ethnography and biopolitics: tracing 'rationalities' of reproduction across the north–south divide. *Anthropology & Medicine, 19*(2), 137–151. https://doi.org/10.1080/13648 470.2012.675050.

Law, J. (1992). Notes on the theory of the actor-network: Ordering, strategy, and heterogeneity. *Systemic Practice, 5*(4), 379–393.

Law, J. (2009). Actor network theory and material semiotics. *The New Blackwell Companion to Social Theory, 3*, 141–158.

Law, J., & Mol, A. (1995). Notes on materiality and sociality. *Sociological Review, 43*(2), 274–294. https://doi.org/10.1111/1467-954X.ep9505171482.

Leigh Star, S. (1999). The ethnography of infrastructure. *American Behavioral Scientist, 43*(3), 377–391. https://doi.org/10.1177/00027649921955326.

Lock, M., & Kaufert, P. A. (2006). *Pragmatic women and body politics*. Cambridge: Cambridge University Press.

MacKenzie, D. (2012). *Missile accuracy: A case study in the social processes of technological change*. In W. E. Bijker, T. P. Hughes, & T. Pinch (Eds.), *The social construction of technological systems: New directions in the sociology and history of technology* (Anniversary ed.). Cambridge, MA: MIT Press.

Mateos Fournier, M. (1964). El incremento demográfico y la planeación familiar. Esterilidad. *Estudios Sobre Esterilidad, 15*(2), 60–64.

McNeil, M., Varcoe, I., & Yearley, S. (1990). *The new reproductive technologies* (pp. 257). No. 28 de explorations in sociology. St. Martin's Press: British Sociological Association. ISBN 0312035993, 9780312035990.

Mejía-Modesto, A. (2007). *Políticas de población, los derechos humanos y la individualización*. México: Gobierno del Estado de México, Consejo Estatal de Población.

Mendoza García, M. E., & Tapia Colocia, G. (2010). Situación demográfica de México 1910–2010. *La Situación Demográfica de México*, 11–24.

Mier y Terán, M. (1991). EL gran cambio demográfico. *Demos, 4*, 4–5.

Mohr, S., & Koch, L. (2016). Transforming social contracts: The social and cultural history of IVF in Denmark. *Reproductive Biomedicine & Society Online, 2*, 88–96. https://doi.org/10.1016/j.rbms.2016.09.001.

Morales Suárez, M. (1999). Princesa 3 pedernal serpiente emplumada simbolo del INPer. *Perinatología y Reproducción Humana, 13*(4), 255–263.

Morales Suárez, M. (2010). Trayectoria del Dr. Eduardo Jurado García (1921–1998). Un acercamiento a su vida y obra. *Perinatología y Reproducción Humana, 24*(3), 207–211.

O'Hara, H. (1944, June). Review of: Doña Eugenesia y Otros Personajes. *American Journal of Public Health, 34*(6), 662.

Olavarría, M. E. (2018). *La gestación para otros. Parentesco, tecnología y poder*. México: Universidad Autónoma Metropolitana and Gedisa.

Pérez-Bustos, T. (2017). "No es sólo una cuestión de lenguaje": lo inaudible de los estudios feministas latino-americanos en el mundo académico anglosajón. *Scientiae Studia*, *15*(1), 59. https://doi.org/10.11606/51678-31662017000100004.

Rodriguez Medina, L. (2019). A geopolitics of bad English. *Tapuya: Latin American Science, Technology and Society*, *2*(1), 1–7. https://doi.org/10.1080/25729861.2019.1558806.

Saade Granados, M. (2004). ¿Quiénes deben procrear? Los médicos eugenistas bajo el signo social (México, 1931–1940). *Cuicuilco, 11*(31), 45–80.

Santacruz-Varela, J. (2010). El aseguramiento de la salud en México y sus tendencias: Del mito al hito. *Revista CONAMED, 15*(4), 195–203.

Simpson, B. (2016). IVF in Sri Lanka: A concise history of regulatory impasse. *Reproductive Biomedicine & Society Online, 2*, 8–15. https://doi.org/10.1016/j.rbms.2016.02.003.

Stern, A. (2002). Madres conscientes y niños normales: la eugencia y el nacionalimiso en México posrevolucionario, 1920–1940. In Laura Cházaro (Ed.), *Medicina, Ciencia y Sociedad en México, Siglo XIX*. Zamora: El Colegio de Michoacán, Universidad Michoacana de San Nicolás de Hidalgo.

Straw, C., Vargas, E., Viera Cherro, M., & Taminini, M. (2016). *Repdoção assistida e relações de gênero na américa latina*. Curitiba: Editora CRV.

Tamanini, M., & Tamanini Andrade, M. T. (2016). As Novas Tecnologias da reproduÇão humana, aspectos do cenário brasleiro, na voz e nas redes dos especialistas. In C. Straw, E. Vargas, M. Viera Cherro, M. Tamanini (Eds.), *Rerpodução Assistida: e relações de género na América liana*. Curitiba, Brasil: CRV.

Thompson, C. (2005). *Making parents: The ontological choreography of reproductive technologies*. Cambridge, MA and London, UK: MIT Press.

Thompson, C. (2016). IVF global histories, USA: Between rock and a marketplace. *Reproductive Biomedicine & Society Online, 2*, 128–135. https://doi.org/10.1016/j.rbms.2016.09.003.

Unnithan-Kumar, M. (Ed.). (2004). *Reproductive agency, medicine and the state: Cultural transformations in childbearing*. Berghahn Books. Retrieved from http://www.jstor.org/stable/j.ctt1btbzpb.

Vallarta-Vázquez, M. (2005). EL consentimiento informado: un derecho reproductivo en México. In M. Torres (Ed.), *Nuevas maternidades y derechos reproductivos* (pp. 239–274). México: El Colegio de México.

Villa Roiz, C. (1997). *Popocatepetl: mitos, ciencia y cultura : un cráter en el tiempo*. Mexico: Plaza y Valdés.

Wahlberg, A. (2018). *Good quality: The routinization of sperm banking in China*. Oakland: University of California Press.

Whittaker, A. (2016). From 'Mung Ming' to 'Baby Gammy': A local history of assisted reproduction in Thailand. *Reproductive Biomedicine & Society Online, 2*, 71–78. https://doi.org/10.1016/j.rbms.2016.05.005.

Zavala de Cosío, M. E. (1992). *Cambios de fecundidad en México y políticas de población.* México: El Colegio de México / Fondo de Cultura Económica.

Zhang, J., Liu, H., Luo, S., Chavez-Badiola, A., Liu, Z., Yang, M. … Huang, T. (2016). First live birth using human oocytes reconstituted by spindle nuclear transfer for mitochondrial DNA mutation causing Leigh syndrome. *Fertility and Sterility, 106*(3), e375–e376. https://doi.org/10.1016/j.fertnstert.2016.08.004.

Zhang, J., Liu, H., Luo, S., Lu, Z., Chávez-Badiola, A., Liu, Z. … Huang, T. (2017). Live birth derived from oocyte spindle transfer to prevent mitochondrial disease. *Reproductive BioMedicine Online, 34*(4), 361–368. https://doi.org/10.1016/j.rbmo.2017.01.013.

Part I

Origin

In 2008, two stories were told, both concerning the origin of the medical business of assisting reproduction in Mexico. The first one was an essay on the history of the Asociación Mexicana de Medicina Reproductiva (Mexican Association of Reproductive Medicine, AMMR), it was commissioned to Efraín Vázquez Benítez, a former president of the association,[1] and published in the inaugural issue of the Revista Mexicana de Medicina de la Reproducción (*Mexican Journal of Reproductive Medicine*). The second was a talk given at the annual conference of the Mexican Gynaecology and Obstetrics Association, by Alberto Kably Ambe, a physician working in the field of AR since the early eighties and also a member of the association. His story was about the first generation of Mexican physicians who started manipulating gametes outside the body as a way of assisting reproduction. Although there were people and places that appeared in both stories, they were very differed stories. What can these stories tell us about the Mexican field of AR and what do they do to the field?

---

[1]He served the 1977–1978 period.

2

Interest in Sterility

Like in other Western countries, during the 1940s, Mexico's medical profession was undergoing a double transformative process: it was professionalising while it was subdividing into specialties. This was part of a larger effort to professionalise many other trades for which president Ávila Camacho signed a law[1] and created the Dirección General de Profesiones (General Ministry of Professions). This ministry would register and oversee the quality the degrees and validated studies done abroad. The law stipulated that, in order to practise a profession legally, you needed to have an official document that certified that you had undergone the necessary training at a recognised institution (Art. 62), and it gave professional associations the authority to set the boundaries of the profession and the acceptable way to carry out such profession (Art. 54) (Soberón, Kumate, & Laguna, 1989).

---

[1] The Ley Reglamentaria del Articulo 5 Constitucional Relativo al Ejercicio de las Profesiones.

© The Author(s) 2020
S. P. González-Santos, *A Portrait of Assisted Reproduction in Mexico*,
https://doi.org/10.1007/978-3-030-23041-8_2

Simultaneously, the epistemic makeup of medical knowledge was favouring specialisation.[2] The medical community (students and professionals) considered that being a specialist gave one a better chance of securing a well-paid job, and it granted the beholder a higher social status. This perception made general professions such as Medico Familiar (Family Doctor or what in some places are called General Practitioner) less attractive. Soon, some of "the disciplines that used to be practiced in conjunction became individual disciplines, for example: ophthalmology and otorhinolaryngology, studies in allergies and in immunology, endocrinology and nutrition, gastroenterology and coloproctology, gynaecology and reproductive biology" (Soberón et al., 1989: 27).

In response to this situation, several medical associations were created during the late 1930s and 1940s, for example, the Mexican Urology Society (in 1936), the Mexican Gynaecology and Obstetrics Association (in 1945),[3] and the Mexican Association of Medical Laboratorists (in 1946).[4] Among the founding members of the

---

[2]A brief note on the different influences the Mexican medicine received during this period: Prior to the 1940s, Mexican medicine was mostly influenced by the French school of medicine; professors taught following the French model of tutors, students used French textbooks, and those who could would spend some time studying in France (in places like Necker hospital, the Hôtel de Dieu, the Broussais, Val de Grace, and the Pasteur Institute in Paris). The first to diverge from this French tradition appeared as early as the mid-1920s; these doctors (among them Salvador Zubirán) chose the USA, particularly Boston and Johns Hopkins. The breakout of the Second World War also contributed to shifting the attention from Europe to the USA, since students could no longer go to France to further their studies and thus turned to the USA. Furthermore, when the USA joined the war, jobs at hospitals became available, thus making the USA an attractive place to study and work. In parallel, as a result of the political turmoil in Spain, during the 1930s and 1940s Mexico received a considerable number of Spanish immigrants, including physicians (approx 500), scientists, and scholars in general (Soberón et al., 1989).

[3]Until 1944, gynaecology and obstetrics were separated fields. Before the unification, gynaecology was a subsidiary field within surgery while obstetrics was a recognised area in and of itself. One of the steps towards this unification was the formation of the Asociación Mexicana de Ginecología y Obstetricia (Mexican Association of Gynaecology and Obstetrics, AMGO), declaring their object of study the understanding of the female reproductive system. The next step was the unification of these two areas in the medical curricula; which happened, in the case of the UNAM, in 1956 and in the hospitals IMSS, ISSSTE, SSA, in 1959.

[4]The Mexican Urology Society began publishing their journal in 1943; both the society and their journal are still active today. The Mexican Association of Gynaecology and Obstetrics started publishing their journal in 1947. Both, the association and the journal, still exist today, but the association has changed its name now to Colegio Mexicano de Ginecología y Obstetricia, COMEGO.

gynaecology association were Carlos D. Guerrero, Alfonso Álvarez Bravo, and Isidro Espinosa de los Reyes, three physicians that played an important part in the configuration of the field of assisted reproduction. During the association's first annual meeting, held in 1949 at the Hotel del Prado in Mexico City, thirty-one men[5] (among them Carlos Guerrero and Álvarez Bravo) gathered to formally establish yet another medical association, this time more specific in interest; an association concerned with infertility and sterility: the Asociación Mexicana para el Estudio de la Esterilidad (The Mexican Association for the Study of Sterility, AMEE). This association is considered the forefather of today's Asociación Mexicana de Medicina Reproductiva (Mexican Association of Reproductive Medicine, AMMR), the professional association that brings together physicians and embryologists engaged in reproductive medicine. As I will describe in this chapter, between 1949 and 2000, the association and its area of interest underwent important epistemic transformations: from the AMEE which was interested in esterilología, to the AMMR which focused on reproductive medicine. Following the story of this transformation is a way of understanding the specialisation and professionalisation process of the Mexican AR system. It is also a way of seeing how the different technologies to assist reproduction and the idea that reproduction can be assisted became disseminated and thus, how the field of assisted reproduction expanded and eventually became an industry. Although the differences between these associations–AMEE and AMMR–and the differences between esterilología and reproductive medicine are considerable, there is something that links them as one and the same association. This continuity has been constructed and maintained through their origin story and the logo of the association (for more on the logo see the end of this chapter).

Unfortunately, traces of the association's history are scarce and scattered since the association has had long periods during which it has not had a journal (or any other means) to record its activities, and

---

[5]An analysis from a gender perspective of the association's history, as well as of the field of assisted reproduction in general, is worthwhile; however, that is not the purpose of this chapter.

it has been inconsistent and unsystematic in the practice of keeping records of what it has done.[6] The way the institutional memory has been kept has varied. During the first twenty years, the association's life was recorded in the association's official publication: *Estudios sobre la Esterilidad* (Sterility Studies),[7] which was regularly published from 1950 to 1970. The only library that has a copy of Sterility Studies is the Dr. Nicolas León Library, UNAM's History of Medicine library located in the city centre, in what was the Spanish Inquisition building and which then became the school of medicine. Sadly, no full collection of the twenty years of the journal publication remains. The only two sets they hold are very damaged, poorly organised, and incomplete; the entire first year is missing, as well as part of the second, and two issues in the early 1960s. Therefore, I only had 54 issues with a total of 481 articles to work with. Through these almost forgotten pages, I was able to identify and remember[8] who these physicians interested in infertility were, how they constructed esteril-ología as their area of interest, and how the association transformed in response to and as part of the changes in the political agenda concerning reproduction and the scientific activities in the areas of gynae-cology, endocrinology, andrology, and biology. For the following

---

[6]At least in recent years, the practice of keeping the material history of the association has been poor. The early material past (journal numbers, proceedings, photographs, articles, etc.) has been lost. The stories that remain have been passed down orally or through publications that appear elsewhere.

[7]The journal was a beauty. Printed in black and white and with hand drawings, it published nine articles on average per issue and printed 1500 copies of each issue. At the beginning, the journal was published four times a year, then, in 1956, it reduced to three, with some years bundling two issues in one. The last issue was published in 1970. Although the journal charged an annual fee ($4 US dollars), it sold issues separately, and sold adverts, it always claimed to be struggling financially. I offered to scan what remains of the journal but the AMMR was not interested. In contrasts, Fertility and Sterility, a journal that began publication around the same time, today is digitally available.

[8]Here I take Haraway's use of remember as an action of reunite, reconnect, reassemble a story, not necessarily as it was once told, but taking elements that had been dismembered (Haraway, 2015).

thirty-eight years (from 1970 to 2008) the association did not have an official journal where to record its social life, instead it was created through rewrites and retellings of the association's history published in journals from other societies, in books, or retold at conferences,[9] which I also take into account. Since 2008, the association has a new journal, *Reproducción*, which I will talk about towards the end of the book. Going through the pages of these journals and books reveals a great deal about the association, the field, the medical profession of the time, and about the country (Callon, Law, & Rip, 1986).

In the following section, I look at one of the most relevant tasks carried out by the AMEE: establishing esterilología as an important area within medicine that they, as a professional association, would develop and practise, and over which they would hold jurisdiction. In order for an occupation to reach the status of profession, or for a medical specialty to be recognised as such, it needs to establish autonomy in three areas: cognitive, normative, and evaluative. The cognitive area, or the epistemic realm, is the body of abstract and specialised knowledge and the set of specific techniques that require prolonged training to acquire and master. The normative dimension refers to the code of ethics that is self-imposed and that justifies the privilege of self-regulation granted to them by society. The evaluative dimension points towards the profession's autonomy to establish their accreditation systems and tend to participate in shaping legislation that pertains to their profession. Robustness in these three dimensions grants the profession a degree of autonomy and prestige within society (Freidson, 1988; Goode, 1960; Larson, 1977). This was AMEE's task during the first decades of its existence.

---

[9]For example, in the ten-year anniversary of the INPer published by Samuel Karchmer in 1993 or the journal Ginecología y Obstetricia de México, which is in very similar conditions to Sterility Studies, no library holds a full set and there are entire numbers missing.

AMEE: The Mexican Association for the Study of Sterility

On the 20th of May 1949 [...] a group of renowned and dynamic physicians gathered to give birth to this association, to formulate a statue that would set its life and work norms with noble, scientific, cultural, and patriotic ends. Its goal [...] has been a strictly scientific objective, to study the problems and divulge the knowledge pertaining to sterility, infertility, and related issues, using the knowledge and practice of the medical, biological, and social sciences. (Valdés la Vallina, 1964)[10]

The men that founded the AMEE came from a variety of backgrounds. Professionally, there were urologists, radiologists, endocrinologists, pathologists, laboratory technicians, and veterinarians, some very well known at the time (Castelazo Ayala, 1959; Castro, 1959).[11] In terms of their institutions, they also came from different working environments; some worked in the public health sector, others in private hospitals (like the Spanish Hospital and the French Hospital). Many of them were also part of other emerging professional associations such as the Mexican Association of Medical Laboratorists, the Mexican Urology Society, and the Mexican Gynaecology and Obstetrics Association. In spite of their many differences, they were all, as said in the above quote, interested in understanding and treating infertility and sterility for "noble, scientific, cultural, and patriotic reasons".

---

[10]"El 20 de Mayo de 1949 [...] un grupo de connotados y dinámicos médicos, se reunieron para gestar esta Asociación, formular con un estatuto sus normas de vida y trabajo, con fines nobles, científicos, culturales y patróticos. Su meta a perseguir [...] ha tenído [...] el objetivo estrictamente científico, de estudiar los problemas y divulgar los conocimeintos acerca de la esterilidad, de la infertilidad y afines, empleando para ello los conocmientos y prácticas de las ciencias médicas, biológicas y sociales" (Valdés la Vallina, 1964).

[11]For example, Francisco Gómez Mont was director of the endocrinology department of the Hospital de Enfermedades de la Nutrición and is considered one of the last century's most prominent endocrinologists in Mexico. Alfonso Álvarez Bravo was the head of the gynaecology service at the Spanish Hospital. Isidro Espinosa de los Reyes, as described in the previous chapter, was the founder of the puericulture movement in México.

As mentioned earlier, the 1940s was a period during which many professionals' associations were formed, struggling to stay alive, and achieve recognition. Being part of an association had economic and practical implications for its members since it costs money and time, thus being part of more than one was a privilege. However, the AMEE was wise in how they administered their overlap with other associations. This overlap allowed physicians to use their economic and practical resources more efficiently by having many of AMEE's annual sessions during other associations' meetings. They did this for almost fifteen years, until they had their first independent annual conference in July 1964. Doing this never meant they lost their identity or their autonomy, as one of its presidents reflects in the association's tenth anniversary "[...] the association has never been seen or used as an instrument by any of the other medical associations, nor has AMEE sought to dominate over them" (Arzac, 1959).

AMEE had a strong conviction of being a national organisation; they would always be aware and attentive to include people from other states in everything they did, for example, schedule sessions in cities other than Mexico City. This effort was particularly notable due to the long tradition of centralising academic, scientific, political, and cultural activities in Mexico City.

From the beginning, AMEE held regular Saturday evening gatherings[12] during which six papers were presented, most were about their clinical practice, others were on their thoughts regarding fundamental issues such as definitions of terms, diagnostic procedures, and their views on infertility. These papers were frequently published in the association's journal Sterility Studies and were considered official papers representing the association's views on the matter. In addition to these commissioned papers, they would also publish translations of articles published in foreign journals (e.g. Slaton Harris, 1951), transcripts of the speeches given by the outgoing and incoming presidents, reports from other conferences, important news concerning the field, or issues concerning the AMEE (such as the

---

[12]These meetings were first held at the Hospital Concepción Béistegui (Regina 8), then during the 1960s they used a conference room at the AMGO's newly acquired building (Baja California 311, Roma), and in 2003 they bought their own office at the World Trade Centre.

association's guidelines). Although the journal was depicted as the official communication medium for the association, it was never viewed—not even by the AMEE members—as a journal of the stature of Fertility and Sterility, which began publication the same year. For example, Carlos D. Guerrero (founding member of AMEE and editor of Sterility Studies) said in an article published in Fertility and Sterility, that Sterility Studies was a "small publication compared to Fertility and Sterility" (Guerrero, 1964). Did they ever aspire to be as important as Fertility and Sterility? There is no way of knowing this. However, it is worth noting that several AMEE presidents complained about the lack of participation of many of its members; they would only contribute financially, paying their monetary fees, but not providing intellectual substance by supplying the journal with articles.

Like the association's membership, the vast majority of the authors published in this journal were men. Of the approximately 480 articles published throughout a 20-year life span, only four were authored or co-authored by women: Edris Rice Wray published twice, first in 1968 which was co-authored with Gorodovsky and then in 1970, Blanca I Conde Manjarrez published in 1967, and Victoria Burciaga in 1968 an article co-authored with Efraín Vázquez. In addition to these, there were a few more papers authored by women, but these were not articles sent to Sterility Studies but only summarised and presented in the section dedicated to the review of the literature (the section was called Resumen de Prensa). These included a paper by Georgeena Seegar Jones in 1951, a paper by Ruth Slaton Harris in 1951, and a paper by Hilde Pfalts in 1956. G. Seegar Jones, together with her husband Howard W Jones, worked in the field of reproductive endocrinology, first at Johns Hopkins and then at the Eastern Virginia Medical School. They are credited with achieving the first IVF birth in the USA in 1981. Several physicians in Mexico were trained at the institute she and her husband opened in Virginia in 1978, The Jones Institute for Reproductive Medicine. Slaton Harris' article was a translation of a paper published in Endocrinology that same year (48(3): 264–272). This paper was part of her Master's thesis presented at the University of Minnesota in 1951. Edris Rice Wray is considered one of the pillars in the research and implementation of the contraceptive pill (together with Margaret Sanger, Katharine McCormick, Gregory Pincus, Luis Miramontes, George Rosenkranz, Carl Djerassi, Rusell Marker, and others). She worked mainly in Puerto

Rico, conducting clinical trials for the pill, and in Mexico, where she established the Clínica Asociación Pro-salud Maternal, the first family planning clinic. Unfortunately, I have not been able to find traces of the two Mexican authors, Blanca I Conde Manjarrez and Victoria Burciaga.

As mentioned above, professionalisation implies the development of a normative infrastructure. This means a code of ethics, a set of rules, a body and process of vigilance and certification, and even a way to discipline those who break the rules. In the early days, the normative framework was influenced by the Catholic worldview. Throughout the journal, both in the content articles and in the association documents that were published there (e.g. end of presidency term report, advertisements for courses, etc.) a Catholic aura can be detected. For example, in his final report as the third outgoing president, J. P. Arzac says: "I am about to finish fulfilling my duties, a weight that at times might seem like a burden, but that at the end it was carried out lightly because it was *bestowed upon us by Christ, Our Lord*" (1956, 7(1): 51) (emphasis added). The presence of Catholic views becomes evident again when Guerrero and Álvarez Bravo decided to publish a transcript of Pope Pius XII address to the attendees of the Second World Congress on Fertility and Sterility, arguing that the Pope's views were of interest to the association since "the majority of the members of our society are Catholic" (1956, 7(3): 59). The reason why I find this worth mentioning is because, in spite of Mexico being a predominantly Catholic country, the relationship between the Mexican State and the Catholic Church has been historically problematic; thus, it was not common to make public claims of ties to the Catholic Church, even when they existed. In both the Constitution of 1857 and of 1917, there are anticlerical articles aimed at diminishing and limiting the power of the Church.[13] The enforcement of these laws has varied, President Plutarco Elias Calles, for example, did try to enforce them as much as

---

[13]Between 1855 and 1863, the period ruled by the liberal Constitution of 1857, the Leyes de Reforma were written and approved. These laws had the purpose of creating a stronger secular state by eliminating the privileges of the Roman Catholic Church, thus diminishing its power. The Constitution of 1917 followed the same path of limiting the Church's privileges and power. This constitution stated that education must be scientific, secular, and fighting against fanaticism. It stated that priests and ministers of religion cannot vote or hold public office; they cannot use religious garb in public, and they set a limit to the number of priest per area.   .

possible, which resulted in the Cristero War (1926–1929). During the following administrations, the tensions between the Church and the State eased, yet never to the point of changing the anticlerical articles in the Constitution. It is in this context that we must think about the implications of presenting the opinions and perspectives of the Church on matters of reproduction in an official scientific journal. While religious ideas could not be used to argue for the legalisation or prohibition of a medical procedure, I wonder how much they did affect the way physicians practised medicine (I suspect that quite a bit).

Finally, it is worth highlighting the words of Castelazo Ayala given during his speech as incoming director of AMEE in 1958: "in our field there are no specialised institutes [...] research is carried out in our own individual spaces, which places a numeric limitation, it has to be done for free, thus it is subject to the limitations imposed by the economy and the work schedule of each and one of us" (1959, *10*(2): 57). In the journal Sterility Studies, there is no evidence that the work done by these particular physicians ever benefited from the national pronatalist agenda, even though they were concerned with curing infertility, and there is no evidence they ever received financial support from any government institution.

After having situated both the association and their journal in the economic, political, and cultural context, I move onto describing the epistemic work they did in terms of conceptualising esterilología as the epistemic foundation of their practice.[14]

Understanding Esterilología Through Its Journal

Esterilología is *"a young specialty which's individuality has just recently been recognised by the medical community"*. (Castelazo Ayala, 1959)

---

[14]This section draws heavily from the articles published in Sterility Studies. As mentioned above, this is a rich journal that has been neglected; there are only two copies left. I hope that readers of this book find this journal as fascinating as I have.

Establishing a medical speciality, or any profession for that matter, is a complex process. Although the formation of the AMEE and the publication of their quarterly journal, Sterility Studies, symbolises the emergence of esterilología as a distinct specialty within gynaecology, they still needed to prove it was relevant. Following the twenty years of journal's publications, we can appreciate how AMEE members worked towards proving its relevance and thus justifying its existence. Throughout these articles, it becomes clear how they constructed sterility as a medical issue complex enough to need a specialised group of professionals to care for it; we can see how their views on marriage, sex, reproduction, and the medical profession shaped their particular way of conceptualising infertility and the diagnostic and therapeutic methods they considered appropriate; and we can also see the process by which the AMEE's code of practice was constructed and transformed. I begin this section by looking at how they justified caring for sterility. Then I describe how they defined marital infertility and conjugal sterility and the epistemic and practical implications of these definitions. I then focus on how they constructed the practice of esterilología and which other solutions they offered at that time. The section closes with an analysis of the type of patients that these physicians saw in their daily practice; this is important because, as mentioned above, it was their own practice what feed their knowledge and ideas of what esterilología was about.

The Importance of Caring for Sterility

One of the first tasks was establishing legitimacy by proving that caring for infertility was important.[15] To do this, they spoke of how devastating infertility can be, pointing out the negative consequences of childlessness

---

[15]Examples of these efforts can be found in several articles published in the Sterility Studies, for example, "Trascendencia de la esterilidad" (Sordo Noriega, 1951); "Plan de organización de la lucha contra la esterilidad conjugal" (Alvarez Bravo, 1952); "Consideraciones sobre la clasificación etio-patológica de la infertilidad" (Santillana, 1952); "Meditaciones sobre la asociación mexicana para el estudio de la esterilidad" (Arzac, 1959).

at all levels: the individual, the couple, the family, and the nation. In his plan to tackle conjugal sterility, Álvarez Bravo (1952) described sterility and infertility as a "problem of great transcendence for both the family and society", because "individuals with children are happier than those without, they live better lives, and they are better members of society" and because "for the Nation, conjugal sterility represents a loss of incalculable human resources", resources that are needed to "amalgamate the past with the future and thus maintain traditions". In addition to this, there was the perception that civilised and prosperous nations had high reproduction rates (see also Sordo Noriega, 1951). They also recognised conjugal sterility as a problem with negative implications for the couple and the individual. As noted by Sordo Noriega in 1951, "procreation is the purpose of marriage [...] children establish the strongest tie between husband and wife", and sterility can lead to the "transgression of moral values", for example, "divorce, adultery, and having children out of wedlock". Álvarez Bravo adds that for women, childlessness is "the in-satisfaction of her maternal desires" which leaves her with "feelings of emptiness" and "psycho-nervous imbalances that endanger the stability of the family", while for men, sterility "mangles his egoism and destroys his aspirations to see his name perpetuated and to educate people that will serve his nation" (Alvarez Bravo, 1952, 3(1): 1).

In addition to proving that sterility was problematic, they had to show why having a professional association occupied with sterility made sense. The arguments they gave invoked clinical, scientific, and educational logics. Regarding the clinical aspects, AMEE hoped that their activity as an association would help increase the number of facilities that offered these healthcare services and to make them affordable to all as part of their right to health care.

Regarding the scientific, they claimed it was important to continue building their specialised knowledge set. As an association, they were engaged with building a robust knowledge set with a basic unified lexicon, a standardised set of diagnostic methods, and a number of acceptable therapeutic procedures (all this makes up the cognitive area mentioned above). AMEE members were particularly preoccupied with unravelling the causes of sterility in order to cure it. They did not want to develop an immediate practical way of overcoming it

(Álvarez Bravo, 1952; Arzac, 1959; Vázquez-Benítez, 2008). Hence, one of their goals was to increase the number and quality of the research being carried out by the individual members of the association and to develop interdisciplinary research projects, in collaboration with physicians across the country and with Latin America that would systematically study infertility and sterility. These robust studies would help produce standardised criteria to interpret test results, establish guidelines to achieve clear diagnosis, and develop effective treatments (Vázquez-Benítez, 2008). One proposal to achieve this was to create a collective database of the clinical cases seen by the AMEE members in their private practice (Murillo & Valdez la Vallina, 1953). However, this database has never been built.

Regarding education, they considered important to improve the education given to the coming generations of physicians in that field. AMEE members were concerned with the education physicians were receiving regarding the treatment of infertility. They believed that these poorly trained physicians were putting women through "unnecessary surgical interventions […] inadequate or even dangerous hormonal treatments without first achieving a correct diagnosis" (Álvarez Bravo, 1952). They were calling attention to the effects of poorly practised obstetric interventions that were leaving women sterile, such as unattended sexually transmitted diseases and infections and the effects of postpartum and post-abortion infections. AMEE members trusted that, by improving the medical training in this field, they were going to be able to improve the quality of the medical attention given to couples facing sterility and infertility. For this purpose, AMEE established formal courses in esterilología. The first of these was offered in 1956 in conjunction with the UNAM, at both private and public hospitals.[16] The programme covered female and male infertility factors (22 lectures on female cases and 11 lectures on male, note the topic imbalance), aspects pertaining to the laboratory, and it held practical sessions for

---

[16]Such as: the Hospital Español, Hospital General, Hospital Juarez, Clinica 2 de Ginecología del IMSS, and the Servicio de Esterilidad del SS.

clinical and surgical areas. These courses are still offered today. A second pedagogical project proposed by AMEE members was to write a collected volume: 45 chapters written by 25 authors, gathering the knowledge and experience accumulated by the association through the years. This book, which had the working title *La Esterilidad y la Infertilidad Humana* would have been published by Prensa Médica Mexicana. However, "a noble and constructive endeavour failed" (Arzac, 1959) because seven months after the due date, only 18 of the 45 chapters had been handed in.

Defining Marital Infertility and Conjugal Sterility

But what exactly was esterilología studying? What was this condition they called "marital infertility" or "conjugal sterility"[17]? In 1952, Manuel Santillana, a physician from Puebla, began his paper on the *Considerations over the Etio-Pathogenic Classification of Infertility* (1952, 3(4): 211–218) by stating that "infertilidad" is not strictly a Spanish word; nonetheless, for lack of a better term in Spanish, he considered appropriate the use of this imported word. Then he defined the condition. First, he distinguished couples who were unable to achieve pregnancy at all, which he called "sterile", from couples who could achieve pregnancy but could not sustain it to term, for whom he used the terms "poor fertility" or "infertile". It is not that he coined these terms; it is just that he officially and formally established that those were the correct terms to use for each of these conditions.[18] He then pointed out the different *manifestations* that infertile couples could display: frequent abortions, foetal death, premature birth, or death of the baby soon after birth. He clarified that these

---

[17]There is mention of marital sterility as early as 1889, in a paper written by Dr. Lier published in the Gaceta Médica de México (est. 1864). There he highlights the importance of studying both women and men; he mentions the use of sperm sample studies and reports that infertility is due to male causes in 44.7% of the cases and to female causes in 55.3% of the cases (in Álvarez Bravo, 1989, 2: 112).

[18]This concept of marital sterility was not exclusive to the Mexican community of physicians, it is also used, for example, by Willis Brown in his Review on Sterility Investigation published in Fertility and Sterility in 1955.

were *manifestations* of a problem (or what we could now call signs), but they are not the *etiopathology* of the problem. Finding the causes for these signs or manifestations was their duty as physicians specialised in esterilología, because they saw the causes as the basis for treatment, and treatment meant curing. Santillana then delimited the scope of infertility and sterility as conditions present in the married couple, hence the use of the terms marital and conjugal. In other words, infertility and sterility were problems *only* married couples could face setting thus the boundaries of who could (a heterosexual couple) and who could not be diagnosed as infertile or sterile (single women and same-sex couples). Furthermore, framing infertility and sterility as "marital" also led members of AMEE to critically address the false yet popular thought "that sterility is always a female problem because it is the woman who gets pregnant and gives life to the creature" (Álvarez Bravo, 1952), and "the erroneous idea that sterility is caused more frequently due to problems or deficiencies of the women, and only exceptionally due to alterations of the man" (Sordo Noriega, 1951, *2*(3): 113–115). Because, said Sordo Noriega, in "daily practice we can see that in almost 33% of cases the causes of sterility are located in the man" (Sordo Noriega, 1951). After all, added Guerrero, "the fertilising capacity corresponds jointly to both the husband and the wife. Sterility can be due to the woman or the man; at the end, the result is marital sterility" (Guerrero, 1954: 29). Consequently, AMEE's guidelines stipulated that "the study of the sterile couple must be directed to both" (Valdés la Vallina, 1954: 15). This particular view justified the frequent use of post-coital tests, which in addition to enabling conformity with the religious prohibition on masturbation, complied with the association's view that infertility could be cured if the right diagnosis was reached.

Nonetheless, attention to male and female factors has never been equal. Just look at Santillana's four possible causes for infertility: (1) alterations in the female genitals, (2) alterations in the general health of the woman, (3) male alterations, and (4) unknown causes. Two sets of causes related specifically to women and only one to male; and if we look at the length, depth, and detail given to each set of alterations, the inequality remains. Santillana spends five pages on female causes while just one paragraph on the male causes. Likewise, throughout the twenty years of publication, the journal published more papers on female factors than on male factors. Within the clinical setting, physicians commented on husband's frequent unwillingness to undergo the diagnostic procedures such as testicular biopsy. And, as I will detail in the following paragraphs, the Catholic Church's opposition to masturbation made studying sperm difficult, the only option being post-coital tests.[19] This de facto lack of study of the male factor might be due to the combination of a moral prohibition on masturbation and coitus interrupts, making the study of sperm more difficult, the lack of cooperation of the male patients in the clinical setting, a wide set belief that reproduction is a female issue, and (male) physicians' poor interest on seeing (fellow) men as (failed) reproductive beings.

Although nowhere in the journal do they openly state that they were trying to comply with the rules set out by the Vatican, they did publish Pope Pius's XII entire sermon where he reflected upon the moral implication of medically assisting reproduction. This sermon was delivered to some of the attendees of the Second World Congress on Fertility and Sterility took place in Naples, Italy, in 1956. As part of the meeting's leisure activities, a trip to Rome was organised giving those interested the opportunity of visiting Pope Pius XII (Tesauro, 1956). The Pope introduced his talk recalling the other occasions when he had addressed some of these same issues with other healthcare professionals, specifically physicians (in 1949),[20] Midwives (in 1951), and

---

[19]If we look at the discipline and clinical activity today, we find that, at the AR clinic, men are mainly studied by the embryologist and laboratory staff (they process, evaluate, and prepare the sperm sample), otherwise they are sent to see the urologist and only in some cases to the andrologist, since only some gynaecologists perform testicular biopsies. Likewise, at conferences presenters rarely talk about the male factors; one example is the 2018 Course on Novelties in Reproduction, a two-day event with 37 presentations of which only one was on male factor and it was from the perspective of the laboratory. This resonates with a comment a male physician made when we were talking about the incorporation of andrology (an area focused on male reproductive endocrinology and physiology) to the field of reproductive medicine, he said: "I studied gynaecology to treat women, not men" (Field Notes, interview with Dr. G, 2007). This comment might be pointing towards a tension in the field's demarcation that might have been in operation since these early days.

[20]In this case specifically concerning artificial fertilisation.

Urologists (in 1953). The purpose of his communication was to present the Church's objection to some of the diagnostic and therapeutic medical practices being used to care for infertility and sterility. As his main justification, he invoked the Church's view on the purpose and limitations of marriage. He stated that, while the purpose of marriage is reproductive, both to gestate and educate the new generations, it is limited to natural reproduction (i.e. as the result of un-intervened coitus) and, although these technologies (e.g. artificial insemination) were tailored to reproduction, they interfere in the natural process. Establishing a civil or religious marriage does not give the couple the right to interfere in the natural and consecrated act of reproduction. In other words, humans do not have the right to intervene, in any way, in the reproductive process, even when the natural (sacred) process fails:

> Artificial fertilisation exceeds the limits of the right that the spouses have acquired by the matrimonial contract, that is, to fully exercise their natural sexual capacity in the natural performance of the matrimonial act. The contract in question does not confer on them the right to artificial fertilisation, since such a right is in no way expressed in the right to the natural conjugal act and cannot be inferred from it. Even less so can it be derived from a right to the "child", as the primary "end" of marriage. The matrimonial contract does not give this right, because its object is not the "child" but the "natural acts" which are capable of generating a new life and which are ordered toward it. It must also be said that artificial fertilization violates the natural law and is contrary to justice and morality. (Pope Pius XII, 1956, 7(2–3): 62)

The emphasis on distinguishing between "the right to a child" and "the right to intercourse to try to conceive a child" is an interesting one. The Pope states that married couples have the right to intercourse but not to the child. This distinction is present again today in the AR clinic's adverts which tend to suggest that you have the right to a child and not just to the act that could lead to conceiving one. For example, in the marketing strategies using the formula "A baby home or your money back", what they guarantee is a child, when what they could be offering a guarantee for would be for the best diagnostic procedure or the highest quality of laboratory procedures.

The Church's official stand regarding masturbation was also very clear: it is absolutely prohibited, regardless the reason to practise it, even if it is to collect sperm for diagnostic purposes:

> masturbation directly procured so as to obtain sperm is not licit, no matter what the purpose of the exam may be [...] even when they are used for grave reasons, which would seem to remove them from culpability: for example, [...] for the medical inspection of the sperm, under the power of the microscope, to determine with which venereal or other kinds of bacterial disease it may be infected; for various types of examinations, from which, it is ordinarily agreed, the semen may be diagnosed by the vitality of the sperm, the presence of components, the number, quality, it shape, strength, and other conditions of that type [...] What has been said up to this point concerning the intrinsic evil of any full use of the generative power outside the natural conjugal act applies in the same way when the acts are of married persons or of unmarried persons, whether the full exercise of the genital organs is done by the man or the woman, or by both parties acting together; whether it is done by manual touches or by the interruption of the conjugal act; for this is always an act contrary to nature and intrinsically evil. (Pope Pius XII, 1956, 7(2–3): 64)

Therefore, in order to study the male factor in reproduction, specifically the characteristics of sperm, the only way around these prohibitions was to perform a post-coital test, where sperm is collected from within the woman's body. The scientific arguments offered in favour of post-coital tests were coupled with the lack of a religious prohibition making thus this procedure acceptable. (I say lack of prohibition because I have not seen any document where the Church openly approves the post-coital test.)

Without the possibility of having extra-corporeal sperm, artificial insemination becomes impossible to practise. However, this was not an issue since many of the specialists from AMEE considered artificial insemination an unacceptable practice in humans mainly because it does not cure the patient, and since the purpose of medicine is to cure, using artificial insemination is betraying medicine's purpose and jeopardising the status of the speciality since it is suggesting that esterilología is incapable of curing. [Although they did recognised the success of artificial insemination as a procedure used in husbandry.] Carlos D. Guerrero wrote in 1953 his clear and emphatic opposition to artificial insemination, on the grounds that it is an immoral and backward technique because it does not solve, i.e. cure, the infertility problem but only bypasses it, which denotes a uselessness of the field's techniques and knowledge: "Yes, for some this technique represents progress, but in my personal view this is not the case. On the contrary, it represents backwardness. More so when the idea is to use a Sperm Bank. These techniques are contrary to the point of view of many and of questionable moral character. Given this, I do not believe the solution to a problem is in artificial insemination, but in correcting the cause that stands in the way of normal insemination. This is why, my point of view regarding this procedure [...] is that, its adoption, only manifests the incompetence of our resources and treatments" (Guerrero 1953, 4(4): 205).

Then, in 1961, Arteaga Elizondo published a paper (previously presented in the IX Gynaecology and Obstetrics meeting held in Monterrey, NL in 1960) where a slight shift in the way artificial insemination was viewed can be detected. In his paper, he concludes that homologous artificial insemination is acceptable, although only for very specific cases because he considers using artificial insemination could emotionally affect the husband. He also recommends having artificial insemination regulated and controlled by a professional body which he suggests should be composed of the members of the AMGO and the members of the AMEE. This is the first time there is mention of the need for regulation. (Note: Assigning jurisdiction to both associations maintains the link between the subspecialty—esterilología—and the specialty—gynaecology. As I will describe in the last section of this chapter, gynaecology has fought against other endocrinology for jurisdiction over ARTs.)

Arteaga Elizondo strongly opposed using donated sperm, not only arguing it was morally wrong and because he saw selling sperm as a mercantile and immoral action (a mercenary behaviour), but because he foresaw that it could have negative effects on men's virility. Using donated sperm could accentuate the inferiority complex felt by husbands when they face male factor infertility and it could affect the relationships between the husband, the wife, and the child. Furthermore, he was particularly against the use of artificial insemination by single women. Although he expresses concern with raising a child without the support and figure of a father, his major objections were regarding the desire a single woman could have for wanting to be a mother; where did this desire come from? He could not imagine how a single woman, with no desires to become a wife, could want to become a mother; how could she be granted the responsibility of motherhood if she has not assumed the responsibility of wifehood? Wanting a child without having first been married, wanting to be a mother without first experiencing being a wife was, according to Arteaga Elizondo, antisocial and anti-sexual. According to him, not only is the purpose of marriage procreation, it is the source of the desire for offspring, and it is the prerequisite to be able to handle motherhood:

> when, for diverse yet inexplicable reasons, a single woman desires a child but does not desire to engage in sexual relations nor get married, in these cases, insemination should not be performed. It is immoral. If using insemination when it is a married woman is already a threat to man's dignity, the faith of marriage, and to human dignity; when it is a single woman using artificial insemination, these same faults are aggravated because in addition she adopts an anti-sexual and antisocial attitude. It is inconceivable that a woman who denies marital responsibility, will acquire an even greater one -maternity- which is greater in biological, social, and legal hierarchy. (Arteaga Elizondo, 1961, 12(2): 63)

In these articles, we see four different objections to the use of ARTs: one regarding the rights to intervene in the reproductive process, another concerned with fulfilling the purpose of a profession, a third about the emotional and psychological well-being of the husband and the child, and the last the problem of separating reproduction from marriage and sex. These concerns reflect the ideas of the time where the purpose of marriage and of sexual intercourse was reproductive, and where male virility was closely linked with his power to produce offspring (the more children he has, the more man he is). A second concern with donated sperm was the commercial aspect, selling sperm was seen as mercantile and immoral; the commodification of body parts, particularly gametes, was still not acceptable.

Esterilología as Practice

How did these Catholic views on marriage, masturbation, and artificial insemination affect the practice of esterilología? Although this is a difficult question to answer, it is an interesting provocation for analysing the articles in Sterility Studies. For example, how often do sperm analysis after masturbation or artificial insemination as a treatment option appear in the journal? None. In contrast, there are over twenty articles mentioning the use of the post-coital tests (or Sims-Hühner test). This diagnostic procedure is not mentioned during Pope Pius XII speech, so it should not go against Catholic prohibitions since it does not resort to masturbation. For the test, the couple has unprotected intercourse, after which the woman's cervical mucus is collected and tested to asses its appearance, viscosity, and elasticity, as well as the concentration, state, and, quality of sperm present in the mucus. Hence, it studies the sperm by looking at it after it has been deposited in the woman's body, and in theory it is a procedure that does not interfere with sperm reaching the egg, so the possibility of impregnation is not tampered with.

In the journal's articles, the rationale behind supporting the post-coital test is not framed as religious but as scientific, as the only test capable of evaluating the interaction between semen and the vaginal and cervical mucus.: "[through this test] one can asses the behaviours of the zoosperm in the medium in which it has to perform, instead of in a crystal box. In other words, within the ph, viscosity, and general state of the cervical mucus and other vaginal secretions. In this mixture of male and female secretions, one can asses the morphology of the male cell, concentration, and motility" (Valdés la Vallina, 1954, 5(1): 18).

In addition to the post-coital test, during the first dozen years of the journal, Sterility Studies, authors discussed diagnostic tests used to visualise the female reproductive organs, for example, the Rubin test (used to see if the fallopian tubes are occluded by insufflation, introducing carbon dioxide into the uterus and through the fallopian tubes), the hysterosalpingography (which is a radiologic procedure to evaluate the uterine cavity and the fallopian tubes), the colposcopy (a visualising technique that can also be used to obtain a biopsy), the hysteroscopy (using a camera, or hysteroscope, to examine the inside of the uterus), and the historiography. A few others focused on cytological evaluation. Regarding the evaluation of male factors, the biopsy of testicles was regarded as an important tool to achieve a full male diagnosis, although there were only a few articles referring to this procedure in the journal (Valdés la Vallina, 1954). Regarding therapeutic procedures, a few articles mentioned the use of surgical procedures although they recognised that these procedures had been quite unsuccessful "we should be ashamed [...] of the few real successes achieved through tubal surgery. Nonetheless, the use of these procedures still increases" (Guerrero, 1953, 4(4): 203). They also employed nutritional therapies in cases of poor nutrition and the use of antibiotics when the cause was an infection. One other procedure they used for several years was applying radiotherapy to the ovaries and the hypophysis. As hope for future developments, Guerrero mentions endometrial transplant and a bank of fallopian tubes for easy reconstruction. It is interesting that, a few lines below in the same article, he criticises the idea of sperm banks. All of these procedures require specialised training, many of which pertain to surgery. As I will explore in greater detail towards the end of part one of the book, the presence of surgery within the toolkit of assisted reproduction has been important when it comes to territorial or jurisdictional disputes between the different medical areas involved in the study and treatment of infertility.

Although these physicians frequently highlighted the importance of studying both men and women when in the process of diagnosing cases of infertility, when it came to the techniques they had to actually study male infertility, these were very poor. Likewise, when reporting on experimental methods, the vast majority of these were focused on women's bodies and processes, rarely on men. Andrology, as the medical area responsible for studying male bodies as reproductive bodies, was never mentioned in the journal.

Other Solutions

A final aspect worth highlighting of these articles is adoption as a valid and suggested solution for involuntary childlessness. (An option that is not offered by the Pope during his speech). Guerrero, in 1954, mentioned that they, as specialist in esterilología—being very much aware of and concerned with the emotional aspects of childlessness—have the authority to recommend adoption as an option for forming a family. Arteaga Elizondo, for example, states that adoption should always be on the table for "we consider highly noble those childless marriages who resolve their situation through the legal adoption since it is a less egotistic, more altruistic option, giving their love, bread, and home to a child, who by all means, is their child. This child will not be an object but the objective towards which they will target their parental energy" (Arteaga Elizondo, 1961, *12*(2): 65). González Gutierrez offers a detailed description of the adoption process in Mexico; he points out that few people are familiar with this process, that there is a need for more formal institutions that can offer organised and legal adoptions, that the younger child is when adopted, the better the outcome, and that—if the adoption process is done properly—"the terrible uncertainty of the hereditary factor will become an almost insignificant matter when related to the nurturing and education given by the adoptive parents if they are a well adjusted and emotionally mature marriage" (González-Gutierrez, 1962, *13*(3): 145). Unfortunately, this open and positive attitude towards adoption is lacking today. As I will describe in the second part of the book, in the public spaces where infertility, involuntary childlessness, and assisting reproduction are talked about (the press, radio, television, trade show, etc.) there is never mention of adoption. During my fieldwork, many women and men spoke about their interest in adoption, but worried that they would not be eligible for a child, due to their economic or marital status, and that their families would reject the idea. In general, it was clear that adoption was less usable than assisted reproduction, that there was less information available and accessible, that it was still enveloped in misleading information.

The Patients

Thinking about the users of esterilología is important because they are key actors in the formation of the field. Physicians develop their knowledge and experience based on the patients they care for in their practice. As mentioned above, members of AMEE worked at both private hospitals (e.g. Hospital Español) and within the public sector (e.g. SS, IMSS, ISSSTE). This actually offered them the opportunity to study a variety of patients. Within this varied population in terms of income and sociocultural background, they found an equally diverse set of causes for infertility: from poor eating and hygiene habits, alcoholism and other substance abuse (e.g. opium and tobacco), sexually transmitted infections (syphilis, gonorrhoea, and tuberculosis), and work-related issues (mainly in men working in the manufacturing sector), to a lack of knowledge regarding intercourse, the negative side effects of ill-practised abortions (which were illegal), delaying marriage, and the use of contraception (which they saw as causes for endometriosis) (Álvarez Bravo, 1952).[21] Some of these concerns echo those held by the eugenicists and puericulturists of the time and have remained important topics still today.

Physicians talked about "poor education" (a nebulous category) as an important factor leading to infertility and to an unsuccessful treatment. This "poor education" was described as either a lack of information or misleading information. They saw it as an obstacle for patients to be able to follow through the diagnostic and therapeutic procedures, for example, when asked to keep a record of their temperature, or as the cause for many of the misunderstandings people had regarding intercourse. Physicians commented that people's ideas about intercourse were full of taboos induced by the educators and the couple's parents, and which they considered an important element resulting in infertility

---

[21]Establishing a causal relationship between postponing pregnancy and the use of contraception with infertility is still a common argument today. The causal relation is sometimes framed as "the effects of the hormones in contraception make you infertile", or "delaying pregnancy mans promiscuity and thus sexually transmitted diseases", and finally "delaying pregnancy makes for an older woman with a diminished fertility".

(Gallo, 1960; Castelazo Ayala, 1960; Climiero-Lafortet, 1952; Lozano, 1952; Weisman, 1953). Santiago Ramírez (1962), a physician and noted psychoanalysts, wrote a book where he talked about the cultural factors involved in infertility. He pointed out that in Mexican culture sexual satisfaction and reproductive satisfaction were seen as antagonistic. Like him, other physicians also related mutual sexual satisfaction and the success of reproduction. In his work, Santiago Ramírez he also mentioned how women from affluent economic sectors were using contraceptives, were breastfeeding for shorter periods of time, and were participating in economic, cultural, and social areas much more than before and more than their counterparts in other economic segments. He describes these behaviours as "formas sociales anglosajoas"—Anglo-Saxon social forms (1960, *9*(3): 149). According to several AMEE members, prenuptial consultations would be a way of dealing with the "poor understanding" of the sexual act which was a cause for infertility, a cause they characterised as the "sexual factor" (Murillo & Valdés la Vallina, 1953; Castelazo Ayala, 1960). Apparently, AMEE physicians did not feel their profession threatened by assigning causality to a non-biological element; it did not result in the loss of authority or jurisdiction over the reproductive area. One reason for this could be that physicians also had jurisdiction over the marriage licence and the prenuptial talk. As mentioned in the previous chapter, as part of the regimes of reproduction put in place in the post-revolution period, couples had to present a medical certificate to the judge in order to get married. This certificate gave testimony that the couple was healthy and that they had spoken to the physician, who in the best of cases would have given them prenuptial instructions. Castelazo Ayala saw these talks as opportunities to give the couple information about the anatomy and functioning of the reproductive organs and the sexual act. Furthermore, puericulture had already assigned physicians and nurses the authority over hygienic matters, which included pre-, peri-, and post-natal care of both the mother and the child.[22]

---

[22]See the previous chapter for a full discussion.

Finally, there were three practical elements that physicians reported as problematic for their patient population: First, economic factors such as the cost of the procedures; second, the long-distance patients had to travel to get to the medical service; and third, the bureaucracies that the public services demands. Although these practical issues affect the patients from the public sector more, they were also reported in cases of private practice. Furthermore, these are aspects that still interfere with patients' diagnostic and therapeutic procedures today (see González-Santos, 2011, 2016).

Summary

I have described Mexico's reproductive agenda from the 1920s to the 1970s, a period during which the country was engaged in a modernising project aimed at becoming an industrialised and scientific country. Part of this project was the effort of professionalising and specialising medicine. This was done through, for example, the establishment of professional associations focused on specific specialities and the new requirement of a certificate to practise. During this time, AMEE, this all-male medical association concerned with sterility, was formed (in 1949). The ideas held by AMEE were, for the most part, congruent with the official pronatalist policies of the time. During this first stage, roughly covering from 1949 to 1966, the association established itself, began the publication of its journal, and initiated regular academic meetings to present papers. These efforts were aimed at justifying sterility as a condition worth studying and treating, of defining martial infertility and conjugal sterility as medical conditions, delimiting the care for involuntary childlessness to heterosexual married couples, and designated which were the acceptable diagnostic and therapeutic procedures. Through the articles published in the journal and the presentations given at professional meetings, AMEE was able to establish a relatively standardised lexicon, a common set of diagnostic and therapeutic tools (e.g. post-coital tests), a purpose (e.g. curing infertility), and a set of moral guidelines (e.g. disapproval of artificial insemination and of single women seeking pregnancy). All important elements for the construction of a discipline and for a field to consolidate and professionalise as a medical specialty. As part of these efforts was the participation of AMEE's members within international associations also concerned with infertility and sterility, for example, in 1957 Carlos D. Guerrero was Vice-President of the International Fertility Association.

Details of a National Portrait

In the many versions told about AMEE's story, nobody has ever made any reference to the association's logo. It has lived the association's changes of perspectives and name adjustments made throughout the past seventy years remaining almost the same. It has worked as an unmentioned almost unaltered link between 1949 and today.

1949–1993. In the beginning, it was a round logo, but not encircled; the perimeter was not drawn. The map of Mexico laid on the bottom of the circle, too big to fit so the Baja California peninsula was left out. Two hands were over the map. The left hand was scratching the middle of the country, preparing it so it can receive the seeds, it was ploughing the land to sow. The right hand held the seeds, letting them fall on to the land. On the top of the circle, the initials A.M.E.E. Between the A.M. and the E.E. a symbol: two snakes coiled around a winged staff, the symbol of Hermes. Although in recent times it has been associated with Western Medicine, in the past it was mostly associated with cunning machinations, trade, eloquence, negotiation, alchemy, wisdom, even thievery and lying (Shetty, Shetty, & Dsouza, 2014). When, in 1966, the name of the association changed, the logo remained the same with the exception of the initials, now being AMEFRH, with Hermes' symbol between the E and the F.

1993. The association changed its name again and updated its logo. It was still a round logo, now encircled. The country was still missing its peninsula, one hand was still ploughing and the other was still sowing. The initials were now AMMR, but between the two Ms a new symbol: the Rod of Asclepius, a serpent-entwined rod; this time the symbol associated with healing since classical times.

2011–2013. For a little over 24 months, the association had different logo. It was still a round logo but now with three concentric circles. The outer one was outlined in a thick red line and white in the centre, with the name of the association fully written on the upper perimeter, from left to right. The second one was a lime green outline that fuses with the third one that was olive green. The third circle was solid. Inside the circle was the full map of Mexico floating over six balls. Given the context, it would not be odd to read this olive green circle with six balls inside as a six-cell (blastomeres) embryo. [However, a six-cell embryo is frequently considered an indication of an asynchronous development, which is not considered desirable (Prados, Debrock, Lemmen, & Agerholm, 2012)]. The hands are not there. This new logo was moving away from the hand sowing, away from the focus on the nation's population and focusing on a more microscopic level, the embryo, and Mexican contemporary biotechnology.

2014–2019. The association goes back to the older version, but with some mild changes. A circular logo, outlined, with the name of the association is written on the upper perimeter, the year of its foundation on the bottom. In the centre, the full map of Mexico with its belly gashed a nearly visible cut. Hovering on top of the country the two hands; the left one has its palm facing down, with its fingers fiddling the cut, ploughing the land and the right one has its palm facing up and is holding seeds that are falling into the cut, sowing. A flow of seeds is connecting the two hands, seeds falling into the land. Two hands working together to fertilise the country. Over the country and fertilising hands is the name of the association AMMR with the Rod of Asclepius.

References

Álvarez Bravo, A. (1952). Plan de Organización de la lucha contra la esterilidad conyugal. *Estudios sobre Esterilidad, 3*(1), 1.

Arteaga Elizondo, O. (1961, Mayo–Agosto). Consideraciones Psicológicas, morales y jurídicas sobre la inseminación artificial humana. *Estudios Sobre Esterilidad, 7*(2), 59.

Arzac, P. (1956, Enero–Abril). Infrome del Presidente de la Asociación Mexicana de Estudios de la Esterilidad correspondiente al periodo 1953–1955. *Estudios sobre Esterilidad, VII*(1), 49.

Arzac, J. P. (1959). Meditaciones sobre la Asociación Mexicana para Estudios de la Esterilidad. *Estudios Sobre Esterilidad, 10*(2), 147.

Callon, M., Law, J., & Rip, A. (Eds.). (1986). *Mapping the dynamics of science and technology.* London: Palgrave Macmillan. https://doi.org/10.1007/978-1-349-07408-2.

Castelazo Ayala, L. (1959, Mayo–Agosto). Discurso del Dr. Castelazo Ayala. Presidente entrante de la AMEE. *Estudios Sobre Esterilidad, 10*(2), 5.

Castelazo Ayala, L. (1960). Importancia de la instrucción prenupcial. *Estudios sobre Esterilidad, 11*(3), 142.

Castro, E. (1959, Mayo–Agosto). UN urólogo en deuda con la Asociación Mexicana para Estudios sobre la Esterilidad. *Estudios Sobre Esterilidad, 10*(2).

Colmeiro-Lafortet, C. (1952, Abril). La falta de orgasmo como causa de esterilidad. *Estudios sobre Esterilidad, III*(2), 57.

Freidson, E. (1988). *Profession of medicine: A study of the sociology of applied knowledge.* Chicago: University of Chicago Press.

Gallo, D. (1960). Frigidez, Orgasmo y Esterilidad. *Estudios sobre Esterilidad, 11*(3), 151.

González-Gutierrez, J. T. (1962). La adopción en matrimonios definitivamente estériles. *Estudios sobre Esterilidad, 13*(3), 137–146.

González-Santos, S. P. (2011). Space, structure and social dynamics within the clinical setting: Two case studies of assisted reproduction in Mexico City. *Health & Place, 17*(1), 166–174. https://doi.org/10.1016/j.healthplace.2010.09.013.

González-Santos, S. P. (2016). Peregrinar: el ritual de la reproducción asistida. In C. Straw, E. Vargas, M. Viera Cherro, & M. Tamanini (Eds.), *Rerpodução Assistida: e relações de género na América liana.* Curitiba, Braisl: Ed. CRV.

Goode, W. J. (1960). The profession: Reports and opinion. Encroachment, charlatanism, and the emerging professions: Psychology medicine and sociology. *American Sociological Review, 25*(6), 902–914. Retrieved from http://www.jstor.org/stable/2089988.

Guerrero, D. C. (1953). Progesos y Problemas a Resolver en la Esterilidad. *Estudios Sobre Esterilidad, 4*(4), 198.

Guerrero, D. C. (1954). Aspectos Sociales de la Esterilidad. *Estudios Sobre Esterilidad, 5*(1), 29.

Guerrero, D. C. (1964). Management of the infertile couple: 25 years' experience. *Fertility and Sterility, 15*(5), 534–542. https://doi.org/10.1016/S0015-0282(16)35348-1.

Haraway, D. J. (2015). *Simians, cyborgs, and women: The reinvention of nature.* New York: Routledge.

Larson, M. S. (1977). *The rise of professionalism: A sociological analysis.* Berkeley: University of California Press.

Lozano, J. L. (1952, Julio). EL factor psico-somático en la infertilidad de la mujer. *Estudios Sobre Esterilidad, III*(3), 153.

Murillo, E. G., & Valdez la Vallina, F. (1953, Abril). Factores etiológicos de esterilidada en 100 parejas. *Estudios sobre Esterilidad, IV*(2), 90.

Pope Pius XII. (1956). Discurso de S.S. Pio XII a los Miembros del Segundo Congreso Mundial de Fertilidad y Esterilidad. *Estudios Sobre Esterilidad, 7*(2–3).

Prados, F. J., Debrock, S., Lemmen, J. G., & Agerholm, I. (2012). The cleavage stage embryo. *Human Reproduction, 27*(suppl 1), i50–i71. https://doi.org/10.1093/humrep/des224.

Ramirez, S. (1960, Septiembre–Diciembre). Factores Culturales en la Esterilidad e Infertilidad. *Estudios Sobre Esterilidad, XI*(3), 146.

Ramírez, C. S. (1962). *Esterilidad y fruto: Psicología de la función procreativa.* México: Pax-México.

Santillana, M. (1952, Octubre). Consideraciones sobre las calisficación etio-patogéncia de la infertilidad. *Estudios sobre Esterilidad, III*(4), 211. AMEE.

Shetty, A., Shetty, S., & Dsouza, O. (2014). Medical symbols in practice: Myths vs reality. *Journal of Clinical and Diagnostic Research: JCDR, 8*(8), PC12–PC14. https://doi.org/10.7860/jcdr/2014/10029.4730.

Soberón, G., Kumate, J., Laguna, J. (compliladores). C. Valdés (coordinador). (1989). *La Salud en México: Testimonios 1988. Especialidades Médicas en México* (Tomo IV. Vols. 1 and 2). Secretaría de Salud. INSP. El Colegio Nacional. FCE: Biblioteca de la Salud.

Sordo Noriega, A. (1951). Trascendencia de la Esterilidad. *Estudios Sobre Esterilidad, 3*(3), 113–115.

Tesauro, G. (1956). Conference programme. In *International Fertility Association.* Naples.

Valdés la Vallina, F., (1954). *Orientación internacional actual sobre esterilidad masculina, V*(1), 14–23.

Valdés la Vallina, F. (1964, Mayo–Agosto). Primera Reunión Anual de la AMEE Discurso del Presidente. *Estudios sobre Esterilidad, XV*(2), 56.

Vázquez-Benítez, E. (2008, Julio–Septiembre). La Asociación Mexicana de Medicina de la Reproducción 1949–2008. *Revista Mexicana de Medicina de la Reproducción, 1*(1), 3–14.

Weisman, A. (1953, Abril). La infertilidad marital debida a problemas emocionales no orgánicos. *Estudios sobre Esterilidad, IV*(2), 85.

3

Managing Reproduction

A very unusual paper was published in the April 1953 issue of Sterility Studies. The author was Abraham Stone, a New York-based urologist involved in the research and promotion of contraception, working closely with Margaret Sanger and with the World Health Organisation (Baker, 2012; Stone, 1953). The paper's title was "Fertility, Sterility, and Population Problems" and its main argument, which was what made it atypical for this journal, was a call to tackle the imminent "population problem": a rapid increase in the total population in Asian and Latin American countries which caused an imbalance "between industrial production and human reproduction" (Stone, 1953, 4(2): 66). According to Stone, the "problem" was the result of a sustained birth rate, a decreasing death rate, and a stagnant production of food. If part of the problem was due to a reduction in death rates as a result of better perinatal care, the use of antibiotics, and the successful control of some epidemics, then it could be said that the post-revolution project concerning health and hygiene was successful. The puericulture practices put in place by people like Isidro Espinosa de los Reyes had fructified. So why was this a problem? For whom was this a problem?

© The Author(s) 2020
S. P. González-Santos, *A Portrait of Assisted Reproduction in Mexico*,
https://doi.org/10.1007/978-3-030-23041-8_3

In 1959, during the midst of the Cold War, General William Draper from the Eisenhower administration, called his government into action to control the rapid population growth that was happening in "less developed countries" because it was threatening the expansion of capitalism, the American way of life (based on high consumption), and life on the planet (Murphy, 2017: 36). His report aided the founding of the US Agency for International Development (USAID) as an arm of the government focused on controlling population growth in these countries and favoured the consolidation of organisations such as the Population Council (1952), the Population Crisis Committee (1965), and the Association for Voluntary Sterilisation (1965). A few years later, at the 42nd conference of the Milbank Memorial Fund, a report on the state of the current research on fertility and family planning in Latin America concludes that "training programs in physiology of reproduction in Latin America should receive full support from concerned agencies and institutions to strengthen them and, if possible, coordinate them at an international level" (Tatum & Delgado-Garcia, 1968: 154).

By the mid-1960s, a new reproductive discourse, like the one presented by Stone in 1953, was gaining force. This discourse claimed that certain countries, particularly Latin American and Asian countries, had uncontrolled growing populations and lacked the infrastructure to care for it. This unbalance, they feared, would affect the global well-being, particularly the USA. These international discourses materialised locally in funding for basic biomedical research on the biological bases of reproduction and on the perception and social implications of managing reproduction. Drawing on books and articles published in the journal Sterility Studies as well as in other journals, in this chapter I will describe how this particular discourse on the "population problem" and the "solution-as managing reproduction" made its way into the AMEE to the point of its transformation. The period this chapter covers (1970–2000) is much less documented than the previous since, during this period, the association had no official journal to keep the memory of its story.

Shifting

In 1968, Paul R. Ehrlich published his best-selling book *The Population Bomb*, in it he argued that, if the growth of the population is not limited, there will be mass starvation and other crisis as a result of overpopulation. This argument located the overpopulation problem particularly within Latin American and Asian countries and it ignited fear that this situation would lead to catastrophic scenarios that could end in wars (de Barbieri, 2000). For example, the USA was concerned with the population growth in Mexico because they considered that "overpopulation and economic conditions in Mexico have helped spur a huge flow of illegal immigrants into the United States" in search of work (Reinhold, 1979); being in the middle of the Cold War, they also feared that socialist and communist ideals would spread across Latin America following the Cuban Revolution. As noted by Murphy, "the problem of population became one of surplus people in 'developing countries' [...] Population was a bomb, in which too many of the wrong kinds of life threatened to destroy economies and the world" (2017: 136). The solution suggested by Stone (and other North American and European demographers and politicians) was the implementation, in these countries and populations, of family planning campaigns that would educate people in reproductive matters, giving them tools to both reduce the number of children they had and space their births further apart. Stone's was the first paper on the "population crisis" and the family planning project to appear in this journal; it was never commented on or replied to in the journal.

Ten years later, in 1964, AMEE decided to revisit the issue. Ignoring it was becoming difficult because, as I will describe in more detail in further sections, by the 1960s there were already research projects studying contraception in Mexico, most of which were funded by foreign organisations (such as the Ford Foundation and the Rockefeller Foundation). So, as part of their first annual meeting (held in Tehuacán Puebla), they organised a round table to discuss the "Demographic increase and family planning". They invited seven speakers to share their views on Mexico's situation regarding the alleged population crisis being spoken of at international forums. These speakers where: the Secretaries of Education (Enrique Rabell[1]), Hydraulic Resources (Humberto Puebla), Agriculture and Husbandry (Joaquín Loredo), and of the Social Security Agency (Pedro Daniel Martínez); a member of the National Institute of Nutrition (Adolfo Chávez Villasana); a faculty professor from UNAM's School of Medicine (Guillermo Corona Uhink); and a member of the International Family Planning Congress (Francisco José Alvarez Lezama). The purpose was to analyse the situation in Mexico, in order to determine if in fact the country was facing a demographic problem, the gravity of it, and its possible solutions. AMEE had no intention of endorsing a particular point of view without first gathering all the necessary information (Mateos Fournier, 1964, *15*(2): 60).

The event was coordinated by Manuel Mateos Fournier, a founding member of the American British Cowdray's hospital medical association and of its journal (Anales Médicos); a member of the Fundación pare Estudios de la Población (Foundation for the Study of Population, FEP); and a researcher on the health-related effects of illegal abortions and the advantages of family planning (1969). During his welcoming speech, Mateos Fournier presented ideas that echoed those presented by Stone a decade earlier but with a twist. He said: "the decrease in mortality rates, the increased survival rate, and the high birth rates have lead to a dangerous unbalance between this and the world's resources, as well as *with the resources each household has*" (emphasis added) (Mateos Fournier 1964, *15*(2): 60). Not only did he mention the effects of the increased population at a national or international level (as other international discourse did), he also considered the effects of more children at the household level. This twist opens the perspective of how to frame the population problem and sets, at least, three possible lines of thought which I will briefly introduce and touch upon in greater detail as the story progresses. Framing number one: certain populations are growing too much and too fast and their countries do not have the resources

---

[1]In the program, it was Celerino Cano who was named as the speaker representing this ministry, but it turn out to be Rabell who gave the talk.

to sustain them, the solution suggested is controlling the reproduction of these specific populations by intervening at the individual
and family level, by promoting (or forcing) the use of contraception.
Framing number two does not question the above framing of the problem—certain populations are growing fast and their countries are not
prepared to provide for them—but does rethink the solution. It suggests implementing policies that favour a more efficient production
and distribution of goods. The third framing does not consider there
to be a population problem, however it does consider important that
individuals and families have the possibility and hability to decide how
many children to have and when to have them (by having access to the
knowledge about reproduction and contraception and to contraceptive methods). Throughout the 1960s and all the way up to the 2010s,
these three framings (and probably others as well) have coexisted and
mixed creating more complex ones. The objective of the round table
organised by the AMEE was precisely to see which sort of framing
they were going to adopt, how they were going to make sense of what
was happening, because it was not evident that the framing and narrative brought from abroad was the only option. There were two main
issues discussed: one concerning the population problem and the other
regarding family planning as the suggested solution.

Thinking About the Population Problem

Physicians were not sure if Mexico had a population problem. To
explore this, they asked their seven guest speakers to address this issue
by talking about: (i) the relationship between population size and poverty, malnutrition, illiteracy, health-care, economic and industrial development; (ii) the effects of mechanisation on employment; (iii) the
availability of water and land for agricultural use; (iv) the logic behind
the import-export practices; and (v) the criteria used by the UN to
qualify Mexico as a underdeveloped country. After listening to these
opinions, some AMEE members were saying that: "We should not fear
a rise in population size if we plan our economy for it" (Alvarez Lezama,
1964, *15*(3): 129).

Regarding the first line of thought, most speakers agreed that Mexico's problems were not directly—or exclusively—related to the increase in population, which in fact some saw as a positive thing, as "the force of the country" (Chavez Villasana, 1964, *15*(3): 123),[2] since it meant an increase in the workforce. Instead of overpopulation being the problem, they identified the lack of proper and efficient distribution of people, goods, and services, as the problems in need of attention.[3] Scholars have pointed out that there was a clear critique to the birth rate decrease solution within other segments of the population (de Márquez, 1984; Welti-Chanes, 2011). In order to understand these points, some context is needed.

Starting in 1940 and up to the end of the 1970s, Mexico went through a period known as the Milagro Mexicano (Mexican Miracle): a thirty-year-sustained development achieved by following a policy of import substitution, promoting the domestic production of goods instead of their import from abroad. The purpose of this was to shift Mexico's economy from a subsistence agricultural economy into an industrialised economy, from a rural nation to an urban one. In this context, workforce was important, and large families were one way of providing workers for the family business, farms, fields, or factories. However, the situation in the rural areas was slightly less miraculous. The amount of land accessible for agriculture and husbandry had not increased at the same rate as the number of campesinos in need of it, so the number of able-bodied Mexicans without employment grew. Many of these (mostly) men headed north to the USA and into the large Mexican cities. Mexico City's population, for example, grew from 7 million in 1970 to 13 million in 1977 (Soto Laveaga, 2007). Cities were becoming overpopulated while rural regions were being left underpopulated. In addition to their precarious working conditions, these rural areas lacked health-care and education services.

---

[2]Alvarez Lezama added to this point that socialist countries also welcome a large population because it means a large work force which in turn make the economy grow. The relationship between population size and socialism was feared by the USA.

[3]Historians have pointed out that there was a clear critique to the birth-rate decrease solution within other segments of the population. For more on this, see Turner (1974), Nagel (1978), Valdés (1980), de Márquez (1984), and Welti-Chanes (2011).

Aware of this situation, many of the speakers suggested improving the conditions in rural areas and thus stopping the migration from the countryside to the cities, instead of reducing their population size—which would leave them with less opportunity to grow economically due to a lack of workforce. In terms of education, the Secretary of Education reported that, although there had been several important actions taken during the past administrations to improve education (e.g. establishing rural elementary schools, the creation of middle schools, the foundation of the National Polytechnic Institute, and the creation and distribution of the free textbook for primary or elementary education), Mexico's expenditure in education was still under most other countries (1.8% annual versus 6.8% in the USSR, 4.3% in North America, 3.2% in Europe, or 2.1% in Latin America) (Rabell, 1964, *15*(2): 80). Alvarez Lezama, from the Family Planning Congress, supported the idea of education as a way to help the population to "naturally turn towards family planning" (1964, *15*(3): 71). Regarding the production, import, and export of goods, although according to the Secretary of Hydraulic Resources (Humberto Puebla) only 15% of the territory could be used for agriculture, the Secretary of Agriculture and Husbandry (Joaquín Loredo) considered that Mexico's alimentary problem was not in terms of a deficient agricultural production, which he assured was enough to feed the nation, but a problem of storage and distribution infrastructure. The lack of this sort of infrastructure meant that the country had to import maize during particular seasons for particular areas because there were no proper storing facilities nor efficient distributing mechanisms to make food reach every corner of the country all year round[4] (these ideas were echoed by Chávez Villasana, a representative of the National Institute of Nutrition).

---

[4]Regarding the export of meat, Loredo said that many Mexicans could not afford to pay for the meat that was produced so they had to export it.

Reducing the number of children does not guarantee a better life [...]
A working class family, with one or two children, does not have the same
emotional protection as a large family; the mental health of the family
can be threatened if the cultural patterns and the social organisation do
not change as well. (Pedro Daniel Martinez, 1964, *15*(2): 101)

The representative of the Secretary of Health quoted above was very
critical of the family planning programmes suggested from abroad.
First, he was cautious with estimating the power of these campaigns,
"they were not"—he said—"the panacea capable of simultaneously
limiting the demographic growth, reducing poverty levels, improving
life, and favouring culture". Second, he warned they should not be "a
means to protect the privileges of a few". Third, he emphasised that
Mexico was a large and diverse country, thus each particular contexts
should be taken into account before thinking about implementing family planning projects developed abroad: "For each cultural, economic,
and demographic situation there should be a tailor made solution". For
many Mexicans, having children had (and still has) a pragmatic component, families are the support network people turn to when it comes to
help during childrearing, illness, and old age (Gutmann, 2009). Fourth,
he pointed out that any project of planned parenthood wrongly pre-
sumed a possibility of evaluating ones own biological, medical, cultural,
and socio-economic characteristics (and that of their own community),
without the possibility of self-determination, the decision of when
and how many children to have would be taken by others or following
ideas and guidelines established by others, an imposition that would be
problematic.

As we can see in all these comments and suggestions, high birth rates
were not seen as the cause of the economic and cultural context (poverty), but the result of them (i.e. poverty caused higher birth rates).
"All cultures", said Rabell, "have planned their reproduction to adjust
to their life style" (1964, *15*(2): 103). Thus, birth rates will decrease
when the economic and cultural wealth of the country is democrat-
ically distributed. Then, and only then, will family planning be the
result of a conscious decision of each individual, and physicians will
therefore be able to provide efficient and personalised means for them

to plan their family size. Sierra García, sub-director of IMSS, assured that "abstinence, the natural way of limiting birth rates, can be achieved by favouring hygienic conditions, the healthy and productive use of leisure time, and by improving the moral and environmental conditions" (1964, *15*(3): 137). The shift of logic presented in these arguments explains family size as an effect of the lack of economic and infrastructural resources at all levels, the individual, the family, and the country— an issue in which the State has a responsibility.

These critiques held similarities with the ideas of Raymond Pearl, an American biologist who considered that it was possible to find ways "to balance quantitative population with national production, bringing biology and state planning together through economy" (Murphy, 2017: 3). According to Pearl, mass death, famine, and overpopulation were entirely avoidable through management, as long as production "progressed" or population was "optimised" (Murphy, 2017: 10). Another alternative to family planning was offered by Sierra García, who suggested the conscious motherhood model (very similar to the idea of conscious motherhood that circulated within the eugenic groups and puericultura), which implied assessing ones hereditary health, economic capacity, and environment, then considering these points to decide the number of children to have. Other contemporary scholars suggested paying closer attention to the relation between a decrease of birth rates and an improvement in access to health, education, and a healthy economic situation. They identified that the reduction in perinatal deaths, achieved during the 1940s and 1950s, were related to better education and improved health-care systems. This in turn allowed people to start thinking about childbearing in a different way: from having many children so at least some would live to having the number of children desired because—most probably—all would live (de Barbieri, 2000; Gutiérrez-Sánchez, 2000; Welti-Chanes, 2011). This shift in attitude might explain the minor reduction in Mexico's fertility rate from 7.3 children per woman in 1958 to 7.2 in 1964 (CONAPO, 1999). Evidence of this shift in attitude is also reflected in the results of a survey conducted in Mexico in 1978 among men and women of middle and lower socio-economic level in rural and urban areas. In it, the respondents distinguished between the size of the families in the

1970s, with 4–8 children, and the size of families in their parents' generation, when couples had over 10 children (Folch-Lyon, de la Macorra, & Bruce Schearer, 1981). Here, we see how improving life conditions helps reduce family size even before the family planning campaigns began.

In general, the ideas presented and discussed at this meeting suggest that these experts saw the population problem, as framed by international agencies, with critical eyes. They recognised there was a problem the country was facing or about to face but understood it differently and thus had other solutions. To further understand their view, I will now turn to how they viewed the family planning programme, the contraceptive solution.

Thinking About the Contraceptive Solution

The second issue analysed during the round table was concerning the solution suggested by some Mexican academics and physicians, and by many foreign international organisations: the use of contraception as a means of family planning. AMEE members were specifically concerned with the psychological and moral effects of controlling reproduction.

"… *frankly immoral*" (Corona Uhink, UNAM School of Medicine)

"*it would be moral to educate our people so they would have less children*" (Laredo, Secretary of Agriculture and Husbandry)

As mentioned earlier, there were two concerns with contraception, one regarding the psychological effects of contraception and the other related to its morality. Guillermo Corona Uhink, a physician from the UNAM with very strong Catholic ideas, considered that the use of contraceptives as "*frankly immoral*" and described the desire to use contraception (as well as abortion and giving children up for adoption) as the result of an alteration of the reproductive impulse (1964, *15*(2): 111), that could lead to mental illness. Like Corona Uhink, Puebla (the Secretary of Hydraulic Resources) viewed any act that halted conception as anti-natural. Others saw the morality and immorality of contraception as depending on the context (Chavez Villasana, López Martinez, and Loredo). Loredo, quoted above, for example, would justify the use of contraception when the family faced economic limitations: "In our country, the highest demographic increases is registered within the poor population. It would be morally better to educate our families so they did not have so many children". Rabell, the Secretary of Education, focusing more on the nation and less on the individual family said: "[...] if a nation needs to increase its population size to develop, then it would be immoral to employ any procedure that would restrict it, but if the size of the population exceeds a limitation, then it is moral to limit its growth" (Rabell, 1964, 15(2): 127). In other words, families should have the number of children they could afford and likewise, nations should have the population size they can care for and need in order to develop as a nation. As we will see further on, this economic talk (affordability) is used during the family planning campaigns in the mid-1970s and 1980s with slogans such as "Pocos hijos para darles mucho" (Few children to give them more).

This brings me to a fourth way of framing the population problem: relating it to consumption. The general framing of the problem was there are too many people and not enough resources. The family planning campaigns developed by some Western countries to be implemented in certain populations was focusing only on one side of the "problem", that of the size of the population. The discussion presented above was suggesting looking also at the other side of the equation: the production and distribution of resources. This other way of viewing the issue is focusing on the interaction between people and resources, in other words: consumption. It is worth highlighting that no family planning campaigns or population policy of this time focused on reducing consumption. Campaigns such as "few children to give them more" and "la familia pequeña vive mejor" ("small families live better lives"), mention the high cost of goods and the problem of income and consumption but they never suggested reducing consumption or increasing income.

One more thing worth highlighting is the place given to the Catholic discourse. During his introduction, Mateos Fournier called attention to the position of the Catholic Church regarding family planning and contraception but, as opposed to what had happened in earlier publications, he politely dismissed it as a point of view to be used as the guiding moral code. Instead, he called for a secular moral, for example during the question and answer segment when he asked the Secretary of Agriculture and Husbandry "if one could be moral without having to adhere to the Catholic moral system". During the first ten years of the journal Sterility Studies, Catholic morals were present and indirectly used as justification for the acceptance and rejection of certain procedures and practices. But by the mid-1960s, although Mexico remained Catholic, and the Catholic Church openly rejected family planning programmes, this did not affect the reception and adoption of the use of contraception since, as some studies suggest, religious practices had little effect on the reproductive decision of Catholic urban couples (Zavala de Cosío, 1992). As mentioned in the previous chapter, the country had gone through a process of secularisation, reducing the power of the Church by, among other things, making civil marriage and birth certificates the only legal document of this sort and eliminating religion from state-funded education. In general, when Catholic morals have made it to policy, they have done so mostly through the pressure of conservative groups within the middle and upper economic segments of the population (de Márquez, 1984).

Promoting and Halting Reproduction: "Two Sides of the Same Coin"

Family planning campaigns are important for understanding the configuration and development of the Mexican AR system because, as stated in the section's title, contraception and assisted reproduction were viewed as two sides of the same coin, a coin that represents both the research and use of hormones and surgery as means to control reproduction, either to reduce it or to promote it. Family planning campaigns worked towards creating the needed material and epistemic infrastructure for assisted reproduction to become acceptable. An infrastructure composed of institutions, people, projects, procedures, and political support, as well as the idea that reproduction, and reproductive problems, can be managed or administered, not only cured.

Up until 1974 Mexico's official policy was pro-natalist (see the 1947 Population Law), a policy created and perpetuated through a series of ideas that, in sum, made it logic to promote increasing the population. These ideas date back to the XIX-century colonisation campaigns aimed at populating the country, particularly the northern territories to protect them from invasions, then to make up for the loss of population due to the Revolution and the epidemics, and then as a way to provide the workforce needed to develop the country (Soto Laveaga, 2007). However, it is important to clarify that there were never policies to actively promote reproduction (such as economic incentives), only to help carry out healthy pregnancies and births (through puericulture). According to de Marquez (1984) during the 1950s and 1960s the de facto policy regarding population was more laissez-faire. Yes, abortion illegal was penalised (since 1931) and contraceptive measures were not sold or promoted in public health-care facilities until the Sanitary Code was changed in the late 1960s, making it legal to prescribed contraceptives under certain conditions. Nonetheless, they were readily available in the market and at the different research sites, because research in contraception was being carried out in different contexts: within the industry (Syntex), in academic institutions (Colegio de México), in government health-care facilities (Hospital de la Mujer), and non-governmental organisations; many of which received funds from abroad, particularly from the USA. So, by the time the foreign pressure was strong enough to influence a change in the official reproductive agenda in 1974, there were already family planning clinics and research groups thinking about and experimenting with different contraceptive methods. Likewise, there were people already using either non-biomedical less invasive methods (i.e. rhythm and withdrawal) or attending the private sector for "modern" family planning methods (i.e. IUD and surgical sterilisations) (CONAPO, 1999; Zavala de Cosío, 1992). Therefore, family planning was received with less opposition than expected (de Márquez, 1984; Segal, 1966; Zavala de Cosío, 1992). In order to better understand the process of developing and implementing the family planning programme, we need to consider, among other things, the activity that the above-mentioned groups and individuals were carrying out before 1974.

The *Asociación Pro-Salud Maternal* (Association in Favour of Maternal Health, APSM), founded in 1958 by Edris Rice-Wray, conducted research on the attitudes people had regarding family planning and the clinical evaluation of oral contraceptives[5] (Lerner, 1967). The *Fundación para Estudios de la Población* (Population Studies Fund, FEPAC), established in 1965 by academics and physicians preoccupied with the rapid increase in population (one of these physicians was precisely Mateo Fournier), had (by 1968) over twenty centres, and in 1969, it became affiliated to the International Planned Parenthood Federation (IPPF) (de Marquez, 1984). Within the private sector, the American British Cowdray hospital initiated, in 1965, a research project to study the psycho-social and medical aspects of fertility and the use of contraceptives. It originally focused on the population of Tacubaya, the working-class neighbourhood where the hospital is located (Lerner, 1967). The academic sector also became involved in the research for the development and implementation of family planning campaigns. In 1964, the Colegio de Mexico founded the Centro de Estudios Demográficos with the financial support from the Ford and Rockefeller foundation and the Mexican Bank, the research carried out there argued for the need to decrease the population size (Caldwell, 2001; Lerner, 1967; Najam, 1996; Soto Laveaga, 2007).[6] Within the medical community, there were physicians from the IMSS and the SSA who were concerned with mother-child health (particularly with the negative effects of early pregnancies, frequent pregnancies, and voluntary abortions). The Centro de Investigaciones en Fertilidad y Esterilidad (Centre for the Research in Fertility and Sterility, CIFE) together with the Hospital de la Mujer, and

---

[5]An active researcher in the contraceptive field, she was the head researcher in the clinical trials to test the pill in Puerto Rico and Mexico.

[6]An example of the work done at this time by Colegio de México is the international meeting they held in 1970 that then became the book Dinámica de la Población de México.

with money from the Ford Foundation and the Population Council, conducted research and training in family planning methods. These projects were coordinated by Alfonso Gutierrez Najar, who at the time was director of the Hospital de la Mujer, and Jorge Martínez Manautou, who was working at Syntex (as I will describe in the following section, they became key figures in the implementation of ARTs in Mexico). They worked towards understanding the way hormones could be used to control reproduction and towards developing more effective surgical ways for permanent contraceptive methods. Regarding surgical procedures, Gutierrez Najar perfected the culdoscopy technique, a surgical procedure developed in the 1940s used to perform a bilateral tubal obstruction (tie the fallopian tubes). Through his technical improvements, he managed to avoid putting patients under full anaesthesia, this implied less complications and faster recovery. This gave him international recognition and led him to establish an international training centre for culdoscopy at Hospital de la Mujer. He and Martínez Manautou, together with H. W. Rudel (from the Population Council) and with R. Aznar, Cortés-Gallegos, and Giner-Velázquez (from the CIFE), were also researching the use of progesterone to control reproduction (Martinez-Manautou et al., 1966, 1967).

Another important activity taking place three decades before the family planning campaigns were set in motion, was the work being done at Syntex first by Russell Marker, then by Rosenkranz, Djerassi, and Miramontes, and even later by Martinez Manautou in the area of hormone synthesis, particularly progesterone, and its use as a contraceptive.[7]

---

[7]Syntex was able to achieve such an important place within the hormone industry worldwide due to several factors. First, Emeric Somlo (a Hungarian lawyer and businessman) and Federico Lehmann's (German chemist and physician) appreciated the advantages that Marker's method of synthesis entailed. When Marker began his work on the syntheses of steroid hormones in the 1930s, it was a very expensive process dominated by the technology developed in Europe (one gram of progesterone could cost up to 80 dollars). This might have been what led him to decided to focus on the source of the compounds, turning his attention away from animals and towards plants, and developing what is now known as the "Marker Degradation". Using plants and his synthesis method brought prices down. Somlo, Lehmann, and Marker founded Syntex in 1944. Second, the high academic standards of the Chemistry school a the UNAM. After a year of

Family Planning Campaigns

When Luis Echeverria took office in 1970 as Mexico's 55th president (since independence), he cancelled all research as well as the educational and clinical activities regarding contraception and family planning carried out at state-run institutions (like the Hospital de la Mujer). During his presidential campaign, Echeverria had used the slogan "poblar es gobernar", to populate is to govern, a direct opposition to the international pressure to control the demographic explosion. An opposition that was not uncommon. As I have shown, speakers at the roundtable organised by AMEE were sceptical of the population discourse. However, during the course of the following four years, perspectives on family planning radically changed.

In 1974, after the International Conference on Population and Development was held in Bucharest, family planning became part of the presidential agenda. This change of attitude resulted in several important political moves made to reinforce the new perspective on

---

having founded Syntex, Marker left. Somlo and Lehmann hired George Rosenkranz, together with a team of Mexican chemists. Soon they were able to develop a similar process and continue the production of hormones. A few years later, in 1949, Carl Djerassi also joined the team followed by a young chemist Luis E. Miramontes. By this stage, Mexico's hormone production was overtaking the monopoly held by Europe and then the USA. As a young chemist, Luis E. Miramontes had been searching for a way to avoid miscarriages when he found a way to avoid conception. In 1951, Djerassi, Miramontes, and Rosenkranz were able to synthesise 19-nor-progesterone, from which Miramontes was then able to produce 19-noretindrona which became the base for the oral contraceptive pill. Due to the importance of this work, Djerassi considers that "Syntex as a company, and Mexico as a country, deserve full credit as the institutional site for the chemical synthesis of an oral contraceptive steroid"; however, this recognition was not given to any of them back then (and still is not fully). Djerassi's statement makes sense if we remember that the American scientist John Rock chose Envoid, the oral contraceptive formula developed by Searle Pharmaceuticals, over the one developed by Syntex (norethindrone), guaranteeing with this that it was the USA and not Mexico who received the credited as the birthplace of the Pill. Third, to the protectionist policies which favoured Syntex in terms of applying extremely high taxes in exportation to foreign labs, delaying or rejecting their permits to exploit the natural resources needed; and finally, because their research interests had extremely practical applications that were in need during the first years of Syntex. For more on the development of the pill, Syntex, and the research on hormones carried out at that time, see Soto Laveaga (2007, 2009) and Léon Olivares (2001).

population growth. For example, the new Ley General de Población was enacted, obliging the state to offer family planning services in public institutions for free and allowing for the advertising and sale of contraceptive methods in public clinics (CONAPO, 1999; Zavala de Cosío, 1992). The Constitution was amended (Article 4) to state that every person has the constitutional right to decide freely, responsibly and with information, about the number and spacing of children they have (CONAPO, 1999).[8] The Consejo Nacional de Población (National Population Council) (CONAPO) was created in 1974 and assigned the task of transforming people's attitudes towards parental roles, the ideal family size, and the use of family planning methods. By the time all this was happening, the attitude of the member's of the government had already changed. In other words, these policies were representing the views of many government members, of many academics, even of certain sectors of the general population.

Only eight years after AMEE's 1966 roundtable, the secretaries of the different branches of government held a different perspective. At the inauguration of CONAPO on 27 March 1974, the speakers, who were also secretaries of the different government ministries, talked about how overpopulation was a direct cause for poverty, illiteracy, rural and urban migration, malnutrition, and plagues (Soto Laveaga, 2007). Instead of resorting to education (as understood in the past through prenuptial talks or basic schooling), they now decided to use publicity and marketing strategies.[9] They turned to the Consejo Nacional de Publicidad (National Publicity Council), a privately funded entity, for help to create campaigns to promote the advantages of having a small family; this was a group comprised of marketing firms and business people.

---

[8]Article 4 "Todos los individuos tienen derecho de decidir, de manera libre, responsable e informada, acerca del número y espaciamiento de sus hijos" Every individual has the right to decide, freely, responsibly and well informed, about the number and spacing of their children. With this, the use of contraceptives becomes part of women's right and part of the Human Rights movement.

[9]This strategy was used in many places. For example, in 1968 Walt Disney together with the Population Council put out short cartoons explaining the problem of overpopulation, how we came to have it, what can be done by using family planning methods, and how we should measure "manhood" in terms of how well he takes care of his children and not on how many he has.

The campaigns promoted by the National Publicity Council were developed by two very successful and renown marketing firms: Noble & Asociados and McCann Erickson-Stanton. They created visual campaigns with posters hung in public spaces such as bus stops and published in the printed press. In both cases, these posters had a slogan or catchy phrase accompanied by an image (a photograph, a drawing, or a cartoon) and a small explanation. Both campaigns sought to promote two messages: having fewer children and improving civic behaviour. In the campaign developed by Noble & Asociados, the image was a family or an individual in a given situation and the phrase "Vámonos haciendo menos…" (let's become less…) followed by a specific message promoting different values and objectives while discouraging others. These could be: "let's become less… macho", "… less submissive" (in case of women), "…less people", or "…less corrupt". The idea was not just to convince people of having fewer children, but to promote gender and civic roles considered desirable "so we can all live better", which was the closing phrase of the formula. The formula created by McCann Erickson-Stanton was "La familia pequeña vive mejor" (small families live better), and in each case, a specific example of why a smaller family lives better was shown both with images and with text. For example, a poster with the image of a basket with vegetables and tortillas said "Thinking about costs? Then think about how with a planned family you can balance your budget adequately. And remember, a small family lives better".[10] Another example was an image of a saucepan with four baby bottles that are being heated and one on the side because it no longer fits in the saucepan, the texts said "Thinking about having more children? Then think about how they will have all they need. And remember, a small family lives better".[11] McCann Erickson-Stanton created a second campaign. As opposed to the previous campaigns, which were targeted to both men and women, this campaign was designed

---

[10]"¿Piensa en el gasto? Entonces piense que con una familia bien planeada se puede balancear adecuadamente su presupuesto. Y recuerde que la familia pequeña vive mejor".

[11]¿Piensa tener más familia? Entonces piense en que todos tengan cuanto necesiten. Y recuerde que la familia pequeña vive mejor".

specifically for women, as can be seen in its catchy phrase: "Señora, usted decide si se embaraza" (Ma'am, you decide if you get pregnant), the image showed a female physician, looking straight into the audience's eyes with her index finger pointing directly at you. According to Soto Laveaga (2007), this campaign was negatively taken by the male audience since they did not appreciate a message that suggested and promoted female reproductive independence. Men considered it was their decision how many children to have, and not their wives (Soto Laveaga, 2007). A third campaign soon came out with the purpose of returning the reproductive decision of family size, if not to the man exclusively, to the couple: "Family Planning: It's a couple's choice". In this case, the image had a couple, the man standing tall next to his wife, firmly holding her and looking at her while she speaks into the microphone of an interviewer out of sight. Although it still was the woman who was speaking, she was being held physically and visually by her husband, returning to him the power he felt he had lost in the previous campaign.

Judging by the images and the texts, these reproductive messages were targeted to the population with economic limitations, since promoting having fewer children was argued under an economic logic: have the children you can afford, and because you can't afford much, have few children.[12] Likewise, these campaigns had a strong emphasis on the individual, on you, your children, and your family as the ones who would benefit from a smaller family. However, as mentioned already, these campaigns did more than discipline reproduction, they also sought to place reproduction (particularly the control of) as a private choice, and they sought to discipline gender and civic roles. In this sense, women were being presented as active responsible individuals, with opinions and ideas, and capable of changing and improving both their life and their family's life. These campaigns aimed at forming the Mexican woman who would have to face the changing global scene.

---

[12]Soto Laveaga (2007), in her very detailed analysis of these campaigns, suggests that these campaigns were targeted to a wider audience. However, her focus when saying this is on the campaigns promoting certain gender and civic roles not reproduction. I agree that these gender and civic messages could have been targeted to a larger audience, but not so the case of the reproductive messages. One instance to support my point is the images used, seldom do you see stereotypes of the economically affluent segment of the population.

Some of these ideas were already present, during the puericulture movement, in the notion of maternidad consciente; however, they acquire more robustness during these campaigns when women are told to stop being submissive and take action and control over their reproductive life (Soto Laveaga, 2007). As we will see in the following sections and chapters, the message of independent women capable of deciding for themselves is used again by the fertility clinics in the advertising campaigns they produced in the 2000s, particularly in the case of campaigns promoting fertility preservation used after 2010.

Campaigns like these were also present in the radio and television. Miguel Sabido, a producer and the vice president of Televisa[13] during the 1970s, was responsible, in coordination with CONAPO, of producing five telenovelas designed to promote the advantages of limiting the number of children to have and spacing their births. The first of these was *Acompáñame* (Come with Me, 1977), then came *Vamos Juntos* (Let's Go Together), Caminemos (Let's Walk), *Nosotros las Mujeres* (Us Women), and *Por Amor* (For Love). Although following the traditional melodramatic structure and cast of a telenovela, these were actually edutainment shows, or educational instruments, with a very clear social agenda and an evaluation package to measure their impact. During the airing of each chapter, CONAPO's telephone numbers were shown so people could call for information about family planning. The success of these soap-operas was measured in terms of, the number of calls registered, the 23% increase in contraceptive sales, and a series of pre- and post-airing questionnaires (Soto Laveaga, 2007).

Soto Laveaga, who has thoroughly researched not only the Mexican family planning campaigns but also the story of barbasco[14] and the development of the contraceptive pill, points out one striking thing

---

[13]Televisa was and is the largest television network in the country. For more on the role of television in Mexican politics and culture, see Beard (2003), Pearson (2005), Orozco Gómez (2006), Tate (2007), González de Bustamante (2009, 2010), Allende (2012), Franco Migues (2012), and Lewkowicz (2014).

[14]Barbasco is the name given to diosgenin-rich Mexican yams, Discorea mexicana and Discorea composita, from which a number of substances (such as steroids) useful in the pharmaceutical industry can be synthesised.

about these state-endorsed family planning campaigns: they present the contraceptive pill as a foreign technology, when in fact it had been a Mexican accomplishment. "None of the hundreds of campaigns and radio and TV spots mentioned that what the government was so actively promoting — contraception in general and oral contraceptives in particular—had been discovered in Mexico, with a Mexican raw material, by a Mexican. [...] When Echeverría inaugurated CONAPO in 1974, less than 20 years after the global achievements of Luis Ernesto Miramontes, he called for Mexican scientists to discover oral contraceptives as the means to help control population growth. By then, however, Miramontes's contribution had already been forgotten" (2007: 29). According to one of his children, Octavio Miramontes, when they called one of Mexico's well-known newspapers to tell them that Luis Miramontes had died, they quoted the cost of an obituary "they were unwilling to publish the death of an unknown person, they were not interested and did not even know who he was" (Paz, 2015).

Soto Laveaga's observation acquires an extra level of interesting when we complement it with de Marquez analysis of the time. She mentions that the government, when planning for these campaigns, was particularly concerned with how the population would take these ideas in terms of Mexico's autonomy as a nation. It was important that these campaigns were presented as rooted in Mexico, not imposed from abroad (particularly not from the USA). Furthermore, it was important to tackle the moral fabric of the Mexican population, primarily the medical community.

By 1978, year in which the first IVF baby was born in the UK, the Coordinación del Plan Nacional de Planificación Familiar (National Family Planning Coordinating Office) was created in Mexico and Martínez Manautou was made head of it. The purpose of this office was to coordinate all family planning programmes in order to meet the goal of drastically reducing population growth by the end of the century. The entire Mexican health-care system was involved. Two highly effective methods were developed: the oferta sistemática (systematic offer) and the establishment of quotas (Pick de Weiss, 1987; Vallarta Vázquez, 2005). The systemic offer meant that women were systematically offered contraceptive methods at every single health-care consultation they went to, even when it was not related to gynaecological issues. The quotas system consisted of a set number of women that had to be either sterilised or incorporated into the family planning programme. Meeting or passing these quotas granted extra income or funds to both the health practitioners and the clinics (Gutmann, 2009). Female surgical sterilisations and the intrauterine device were the methods mostly recommended by the public health service due to their practicality, the perception that they did not require much follow up, and because apparently their abandonment rate was much lower than that of hormonal contraceptives. The majority of the sterilisation procedures were done at public health institutions, usually after giving birth, having a miscarriage, or an abortion (Zavala de Cosío, 1992). Furthermore, in rural areas, women were offered to be taken to the nearest health centre, where they were given a contraceptive method (i.e. intrauterine device, IUD, or sterilisation) and then were taken back to their hometown, all in the same day.

Among the first to adopt family planning methods were women who by the mid-1970s, when the campaigns began, were either in their mid-1930s (women born between 1937 and 1941) and wanted to limit the number of children they had, usually after having the fourth child (Gutiérrez-Sánchez, 2000; Zavala de Cosío, 1989), or were starting their marital life (women born between 1942 and 1946) and wanted to delay pregnancy (Gutiérrez-Sánchez, 2000). This is one of the first clear examples of a growing desire to delay pregnancy, something that becomes central within the discourse of infertility and assisted reproduction.

The year Echeverría took office was the same year the journal Sterility Studies published its last issue. According to some, it was due to economic problems, they said it never was able to sustain itself. It would be almost thirty years until the association began publishing a new journal, the Revista Mexicana de Medicina de la Reproducción. I will come back to this journal and to this period of the association further ahead.

Reproductive Biology as a Research and Academic Programme

As I have detailed, between the 1940s and the mid-1960s, infertility was dealt with through the knowledge and tools that constituted esterilología. Then, during the 1960s, a new set of tools and premises were being developed within a handful of research facilities, which had foreign financial support (Syntex, National Medical Centre, and the National Institute of Nutrition). This new knowledge set was the emerging field of biology of human reproduction, the field that eventually will be the epistemological base for assisted reproduction.

Reproductive biology is a multidisciplinary area that has as its main purpose the study of the function of the human reproductive system in order to diagnose, prevent, and treat its alterations. As a multidisciplinary area, it brings together knowledge from biology, medicine, and the social sciences. Although alterations in the reproductive process were of interest of medicine in general, and other medical specialties in the past, the discourse and fears of overpopulation that emerged by the mid-twentieth century, turned understanding the elements and processes involved in reproduction urgent. The knowledge created and accumulated with this purpose—understanding the reproductive process and its alterations—is what constituted reproductive biology. Interest in reproductive biology gave place to the creation of clinics, services, and departments with reproductive biology as its focus, as well as a professional association: the Academia de Investigación en Biología de la Reproducción (the Academy of Reproductive Biology Research).

Pérez Palacios and Ulloa Aguirre identified three key moments in the background story of reproductive biology in Mexico. First, the research that Eliseo Ramírez was conducting in the 1920s demonstrated that there were changes in the vaginal epithelium in the different stages of the menstrual cycle and pregnancy. These were similar studies and findings, but conducted prior to, those conducted by Papanicolaou. Second, the experimental work conducted in the 1940s by Efrén Pozo and Efraín Pardo, and then by Alfredo Gallegos Cigarroa, with an indigenous plant called zoapatle (which was already mentioned in the Cruz Badiano Codex as having abortive properties) (Miguel Ortega, Arriaga Dávila, Sepúlveda Vildósola, Salamanca Gómez, & Martínez Castuera Gómez, 2018).

Third, the research on the effects of nutrition on hormones and reproduction done by Francisco Gómez Mont[15] (founding member of the AMEE) at the National Institute of Nutrition in the 1950s. Several of the students that worked with Gomez Mont also trained with US-based physicians like Raph I Dorfman and Gregory Pincus, who developed ways to measure hormonal levels as well as ways to regulate reproduction through the administration of hormones, and with Robert Greenblatt[16] and M. B. Lippset, who were developing new ways of treating infertility. Carlos Gual Castro (2000), another key figure in the establishment of the field, considers the birth of reproductive biology to take place between 1965 and 1966, years in which the first two education and research departments focused on reproductive biology were created, one at the Instituto Nacional de Nutrición Salvador Zubirán (National Institute of Nutrition Salvador Zubirán, INNSZ), directed by himself, and at the National Medical Centre, directed by Jorge Martínez Manautou.[17]

The National Nutrition Institute was created in the 1940s as a specialised hospital housing the first hormones laboratory in the country (Hernández Valencia & Saucedo García, 2018). In 1965, after the new department (independent from the department in endocrinology) focused on researching steroid hormones and reproductive pharmacology was established, they opened a clinic to care for reproductive endocrinology, offering family planning services and sterility care, and a training programme. The head of this clinic was Carlos Gual Castro, who studied the use of urinary gonadotropins to induce ovulation as a way of treating infertility. Gregorio Pérez Palacios was head of the academic programme. It was a two-year programme, approved by the

---

[15]Endocrine Disturbances in Chronic Human Malnutrition. (1953). *Vitamins and Hormones*, Vol. 11, pp. 97–132.

[16]Robert Greenblatt (1906–1987) was a Canadian endocrinologist who worked at the Medical College of Georgia. He was involved in the development of the sequential oral contraceptive pill and in the discovery of clomiphene citrate as a drug to induce ovulation.

[17]President Manuel Avila Camacho (1940–1946) developed the National Hospital Plan as a way to increase the number of hospitals and expand their reach to cover larger sectors of the population. Both the National Nutrition Institute and the National Medical Centre were part of this project. For more on the National Medical Centre, see the work of Fajardo Ortiz, a researcher at UNAM who has done extensive work on this institution.

Universidad Autónoma Nacional de México (National Autonomous University of Mexico, UNAM) in 1968 as a medical subspecialty, as a masters programme, and as a doctorate programme. The first class began its training in 1967, and during its first 30 years more than 120 students graduated from their programmes (Gual Castro, 2000; programa de la especialidad).

A couple of years later, in 1969, Jorge Martinez Manautou and Juan Giner created the second programme at the National Medical Centre. The story of the National Medical Centre began in 1945, when the Ministry of Health acquired a 156,000 sqm lot of land to build a network of seven hospitals with a total of 2000 beds, that would not only offer the best possible care for the Mexican population, but also be a place for high quality research and teaching; in other words, it would be the National Medical Centre. In 1961, the IMSS (the Mexican Social Security Institute) bought the medical complex and refurbished it, adding and eliminating services, offices, and buildings. In 1966, Jorge Martinez Manautou—brother of Emilio Martínez Manautou, a member of President Gustavo Díaz Ordaz's cabinet—managed to secure financial support from the Ford Foundation (one million dollars) and political support from President Díaz Ordaz to establish, at the National Medical Centre, a research centre focused primarily on reproductive biology, the Unidad de Investigación Científica (Scientific Research Unit) (Zárate & Basurto-Acevedo, 2013). The purpose of the Scientific Research Unit was to carry out research that would help understand, among other things, the physiology of reproduction to be able to manipulate hormones and manage reproduction with family planning purposes. Martinez Manautou had a history in the field of hormonal contraception since he had worked at Syntex as head of the clinical research unite (Martiñez Manautou, 2013; Zárate, 2015). Together with the research programme, an academic programme was established under the direction of Jorge Martinez Manautou and Alfredo Gallegos.

As mentioned above, during the 1960s and 1970s, promoting and halting reproduction were two sides of the same coin, the coin of the reproductive biology, which was defined by Canto de Cetina as "the science of the transmission of life" (1994: 145). The question asked by reproductive biology was a basic science question, "which are the biological elements involved in reproduction?". This contrasts with esterilología's question, which was being asked a few years before, and was oriented more towards the clinical setting: "how to cure marital infertility or conjugal sterility?". Although esterilología and reproductive biology share an interest in infertility, these two fields also differed, not only in their epistemic and technological aspects but also, as I have detailed above, in a sociopolitical aspect. For example, in terms of who was practising them, where they were practicing it, and how these practices were being financed. These differences are embodied in the new name of the medical association, AMEE changed its name to AMEFRH.

AMEFRH: Mexican Association for the Study of Fertility and Human Reproduction

The continuous development of the specialty through the creation and optimal management of new systems to regulate fertility, as well as through the development of new and complex diagnostic and therapeutic methods for the medical treatment of infertile couple, have conditioned the development of the new subspecialties in contraception, andrology, and infertility. Regarding this last area, the specialist needs to have knowledge and skills in the area of gamete preservation and transference, and in in vitro fertilisation, which probably represents the most complex human reproduction therapeutic. (Pérez Palacios & Ulloa Aguirre, 1989: 235)

Despite the hesitation regarding the existence or not of a population problem in Mexico and with implementing family planning as a solution, the force of these ideas and of the money invested in this project permeated and eventually was adopted by the Association. From 1965 up until 1970, when the last issue of Sterility Studies was published, there was a constant increase in the number of articles concerned with contraception, a reflection of the increase in research projects and clinical practice around family planning. AMEE's 1967–1969 administrative board, for example, featured doctors who were already offering their patients oral contraceptives. Congruently, during the V annual meeting (held in 1968 in Mexico City) there were more speakers addressing contraception than ever before. However, the most important action that reflects this shift took place in 1966 when, under the leadership of Alfredo López de Nava (who was a founding member of AMEE), the association decided to change its name to Asociación Mexicana para el Estudio de la Fertilidad y Reproducción Humana (Mexican Association for the Study of Fertility and Human Reproduction, AMEFRH). During his acceptance speech as the newly elected president of the association, Francisco Durazo Quiroz argued that the new name reflected more accurately the association's activities and interests, which had moved from esterilología, a perspective which focused on infertility as a health problem in need of a cure, to focusing on fertility as a socio-economic problem in need of medical and political (biopolitical) solution, mainly through family planning (1967).

During this period, the academic stature of reproductive biology reached the necessary level to be considered a field worth being incorporated into UNAM's postgraduate medical curricula, initially as a specialisation programme and then as masters and doctoral degree (Gual Castro, 2000). Simultaneously, doctors at the Gynaecology and Obstetrics Hospital No. 1 began studying the effects of immunology on infertility (Tatum & Delgado-García, 1968). Several years later, in 1974, the newly established *Universidad Autónoma Metropolitana* (Metropolitan Autonomous University, UAM) and the National Institute of Nutrition created an undergraduate programme in reproductive biology, and almost two decades later, in 1991, the *La Salle* University, in conjunction with Gutierrez Najar's private fertility clinic, opened its postgraduate degree within their medical faculty. As I will detail in Chapter 4, the INPer opened its course on reproductive biology in 1986, which meant they also had a clinical service where they performed artificial insemination, IVF, and GIFT (Karchmer, 1999).

As mentioned before, the last issue of the Sterility Studies was published in 1970. So, in order to continue tracing the activity of the association and the development of the field, I turn to three books published within a twenty year period (1976, 1987, 1999). In these publications, I found evidence of changes in their epistemic, moral, normative, practical make-up as well as in their self-perception; changes in how they conceived infertility, in how they were diagnosing it (where they still using post-coital test or now accepting masturbation as a way of collecting sperm?), in how they were treating it (were they still seeking to cure it or now accepting to just achieve pregnancy without curing infertility?), and in how they saw themselves and their activity.

The first two books, *Esterilidad e Infertilidad* (1976) and *Manejo de la Pareja Infértil* (1987), were written and published in the period during which the association was called AMEFRH, the family campaigns were at a rise, which meant a great deal or research on reproductive biology was being carried out financed with funds that, in many cases, came from abroad, Mexico's economy was still relatively closed, and high complexity assisted reproductive technologies were still not successfully used in México. The authors of these first two books were primarily physicians doing research at publicly funded research facilities, as opposed to many of the authors in Sterility Studies or the authors of the third book, who were clinical physicians working at hospitals. The third book was published in 1999, once the association had, again, changed its name in 1992 under the administration of Carlos Hinojosa; the new name was Asociación Mexicana de Medicina Reproductiva (Mexican Association of Reproductive Medicine, AMMR).

Two Books

I begin with *Esterilidad e Infertilidad*, published in 1976 (and reprinted in 1986), by Arturo Zárate (working at the Scientific Research Unit of the National Medical Centre), Elias Canales (president of the AMEFRH in 1975–1976), and Carlos MacGregor, in collaboration with five other authors, two of which were also former AMEFRH presidents: Juan Griner (1976–1977) and Jorge Delgado Urdapilleta (1973–1974). The book was dedicated to Luis Castelazo Ayala, also a former president of the association (1959–1961) "who, through his support and encouragement, made possible for the Hospital de Ginecología N1 of the IMSS to offer scientific attention to the infertile couple".[18] It was prologued by Robert Greenblatt and written for students of medicine and physicians in general. There are at least four things that this book does.

First, explicitly through Greenblatt's prologue and in the authors' introduction, and implicitly throughout the chapters, the book positions the study of sterility as a consolidated field and as an established "clinical science" (Zárate, Canales, & MacGregor, 1976). This status has been achieved, they explain, by to the advancements made in the area of physiology of reproduction, particularly regarding the knowledge and clinical experiences gathered around the use of LRH (work done by Andrew Schally, Zárate, Canales and Kastin, among others), clomiphene citrate (work done by Greenblatt among others), and gonadotrophic hormones.[19]

Second, the book presents a shift in how to diagnose and treat infertility due to male factor. Juan Giner, author of the chapter dedicated to this subject, includes the spermiogram (semen analysis) as an acceptable diagnostic procedure explicitly stating that this requires masturbation, and it considers artificial insemination as a therapeutic option, including the use of donated sperm. Both procedures were available before the 1970s, yet they were considered unacceptable by most AMEE members, as was clear in the analysis of the articles in Sterility Studies. Admitting artificial insemination as a therapeutic option signals an important change in how they saw the purpose of their field. Artificial insemination does not cure infertility, although it does helps conceive. Hence, their goal was shifting towards helping couples conceive without having to cure what was causing infertility.

Third, the last chapter is a two-page comment on adoption. The authors say that, while it is the physicians who commonly have to suggest this option to the couple when all other techniques have failed,

---

[18]Sixth president of the AMEE (1959–1961) and the first that was not a founding member, responsible for the work on sterility carried out at the Ginecol of the IMSS, where three of the authors worked.

[19]LRH is the luteinising-releasing hormone, GnRH is the gonadotropin-releasing hormone.

they commonly lack information about the process, its implications, and outcomes. Contrary to what many people think and fear—that families with adopted children commonly have problems—they present studies that indicate the contrary. They claim that the studies they have consulted conclude that, in fact, the majority of the cases of adoption are successful. They suggest the best practice is to undergo a legal closed adoption but always letting the child know she or he was adopted.

What the book does not do is change the unbalanced attention given to male and female reproductive organs and processes. This is visible in the book's structure. Eight out of fourteen chapters are focused on female factors, covering issues on anatomical or physiological alterations in ovulation, ovaries and fallopian tubes, cervix and uterus, and endometrium. Only two are focused on male factors and none with a direct reference to marital sterility.

Ten years later, in 1987, Zárate and MacGregor publish *Manejo de la Pareja Estéril*—a book written in collaboration with seven other authors, three of which were women: Carmen García Mendieta, an academic lawyer from the UNAM, who wrote about the legal aspects of reproduction; Esther Moscona, a psychologist, who wrote about the psychological implications of infertility and adoption; and Martha Medina, a researcher from the Scientific Research Unit of the National Medical Centre, who wrote about the physiology of reproduction. This book was published nine years after the birth of Louise Brown and two years after the first two IVF clinics had been opened in Mexico. Like in the previous book, important shifts in perspective can be detected. Shifts in progress, transitions, this progressive element can first be detected in the book's epigraph: an extract from D. H. Lawrence Lady Chatterley's lover (1928)[20] where Lady Chatterley reflects on the difference between having a child by herself and having a child with someone she loves and lusts, the difference being that in the second case she feels the profound feeling of femininity and the dream of creation. The novel's plot, use of

---

[20]A novel banned in many countries until roughly the 1960s because of its explicit descriptions of sexual activity between a working-class man and an upper-class-married woman and the use of words considered unprintable.

language, and its reputation as profane is at odds with the message of the book, which still presents infertility as a couple's matter, as indicated by its title.

The book frames infertility as a medical, social, and cultural problem faced by 10–20% of the population, a *problem* for which they have *solutions* (Zárate & MacGregor, 1987). This contrasts with the way infertility was referred in the past, as a *disease* with emotional and social consequences that needed to be *cured*. This slight shift is the beginning of a slow yet constant move in which the condition—infertility—is being extracted from the medical paradigm. This move sets the conditions for a wider spectrum of user eligibility; something we see much more clearly in the 2010s when people no longer have to have a clear medical problem (obstructed tubes or low sperm count, for example) to be a valid candidate for assisted reproduction. With a wider or more inclusive criteria, new potential users appear: single individuals, same-sex couples, transgenders, etc. This more inclusive criterion of eligibility has generated new topics of debates in which new lines of arguments appear. For example, debates arguing against discrimination: the Grupo de Información en Reproducción Elegida (GIRE) has argued against what they claim to be discriminatory practice taking place when a health system requires a medical diagnosis in order to be admitted into an AR service (GIRE & Radar 4, 2015). This debate is complicated since the "solution" to involuntary childlessness is still located within the jurisdiction of medicine.

This book also suggests there has been a rise in the number of couples facing infertility and suggest that this rise is due to certain social changes affecting particularly women. Some of the changes they point out are: an increase in the number of women working in jobs that expose them to toxic agents, women postponing pregnancy and thus using contraceptives for long periods of time, and the effects of the sexual liberation, which they describe as "an agent that contributes to a higher exposition to infections" and that, together with "the increase of contraceptive use has lead to promiscuity [...] [and to the] exposition to a greater variety of semen (due to multiple sexual partners) can increase the possibilities of developing antibodies against spermatozoa. This leads to an immunologic sterility with a very difficult solution" (Zárate & MacGregor, 1987: 14).

Another reason for the increase in the number of people arriving to the medical consultation was described following a market logic of offer and demand. They claimed there were more couples coming forth because there were more physicians specialising in this field. This market logic is re-emphasised later on as an alert: "currently the medical practice is immersed in a technological and consumerist chasm, and thus, patients that seek to solve their problem can be prey to exploitation. Therefore, it is important that physicians are well aware of the benefits and risks of these new technologies and do not let themselves be seduced by the talismans and glitter that do not benefit their patients" (Zárate & MacGregor, 1987: 7). In this sense, the induced demand is not only in terms of the patients but also in terms of the physicians: "we cannot leave out the spectacular appearance of the in vitro fertilisation with embryo transfer which started only seven years ago and has already reached a commercial state with ample use, regardless of its high cost and its need for highly qualified specially trained staff"[21] (Zárate & MacGregor, 1987: 17). They mention that there are clinics performing IVF, which is "no longer esoteric" (Zárate & MacGregor, 1987: 89), in the USA, Europe, Australia, and Japan, but never mention the two Mexican clinics active at that time. This omission could be because, although the publication date is 1987, the book could have been written before these two clinics were opened in 1985. They consider this procedure can be used for a variety of cases diagnosed with female infertility where there is "sperm of acceptable quality" (Zárate & MacGregor, 1987: 89–94).

It is interesting to compare what they say about IVF to what they say about artificial insemination. Interestingly, while they say that IVF is spectacular, the authors of this book still consider resorting to artificial insemination to be "a full admission that all other therapeutic measures have failed and that there are no other options" (p. 85). As opposed to IVF, artificial insemination is recommended mostly for cases of male infertility, for example where the sperm sample is inadequate or when

---

[21]They sometimes use the term Fertilización Extracorporea (FEC) or out-of-body-fertilisation to talk about IVF.

there is a problem with impotence or ejaculation. They claim that, in spite of artificial insemination being used every day more, some couples still reject it after hearing its medical, psychological, social, marital, and legal implications. Therefore, they remind the reader of the importance of considering the psychological state of the couple, their religious beliefs, and their desires to become parents when offering this option to couples. They claim that in Mexico, the use of artificial insemination with donor is very uncommon, and when it is used, physicians do not tend to report this due to fears of moral or legal consequences. Interestingly, none of these concerns were brought up when talking about IVF. Could gender issues be related to such different attitudes towards each technique, one being framed for female causes is more acceptable for male physicians while the other, directed to male causes, being less acceptable?

In contrast to the previous book, this one has a chapter on the legal implications of assisted reproduction. The chapter's rationale is based on the fact that, given the case, judges have to resolve disputes, even if there are no specific laws for those cases. With this in mind, the author, Carmen García Mendieta, analyses the General Health Law and the Civil Code to see how the law can be interpreted if kinship assignation problems arise in cases where ARTs are used (e.g. artificial insemination, IVF, post-mortem fecundation, and surrogacy). Under her interpretation, homologous artificial insemination and IVF do not present problems, since in these cases neither the biological nor the martial precepts are broken[22] and Article 466 of the General Health Law states that a married woman needs her husband's consent in order for her to be legally inseminated. Donor artificial insemination, however, can be problematic for the donor because there are no laws to protect his anonymity.

---

[22]It is important to remember the basic premises of legal kinship are both biological and marital, the woman who gives birth is the mother and her husband is the father. Kinship is based on the biological link which is the result of copulation. It is assumed that for every birth there was the previous act of coitus.

Like the previous book, this one also has a chapter on adoption. Within the context where Mexico was portrayed as a country perceived as overpopulated and with considerable economic problems, it is a surprise to see that Esther Moscona, author of this chapter, offers a different panorama. She claims that, because of the accessibility to contraception, the continuous practice of illegal abortions, and the increase in the number of single women willing to keep their children, there is a "drastic reduction in the number of children up for adoption" (p. 111). As a result of this, she detects a rise in the number of foreign adoptions.

Summary

This chapter, traces important shifts in the field of reproduction, changes that continued the process of constructing reproduction as a politically and biologically manageable process. (i) Reproduction (i.e. family size and the space between each child) acquired a particular economic and political hue, reducing family size and increasing the age gap between children was framed as a medical, social, and cultural problem in need of solutions. Large populations made capitalist consumption complicated because the resources are limited. Who was reproducing was also seen as a problem, non-whites were reproducing too much. A few critical perspectives were put forth, perspectives that questioned the problem's framing and solution (reducing reproduction) and instead suggested policies and infrastructures to improve the distribution of resources; however, these ideas were never taken up by the State. The family planning projects, which were concocted abroad, participated in this framing. Although at first the physicians of the AMEE were critical of these family planning projects, eventually the association aligned to the project. This became evident when they decided to change their name to AMEFH to indicate their interest now in contraception. (ii) Biology of human reproduction (or reproductive biology) became consolidated as a medical field. In this chapter, I traced how esterilología, which was concerned specifically with the study and care for infertility, transformed or passed away and how reproductive biology, interested in both halting and promoting reproduction, emerged and became the epistemological filed in charge of managing reproduction. (iii) The purpose of this medical field no longer is to cure a medically identified and typified condition, it has transformed into helping their patients conceive (have a baby) or not conceive (have sex without getting pregnant). (iv) Infertility is considered a problem at a rise due to an increase in the number of specialists treating it and to infertility-inducing behaviours. Women's lifestyle now includes work, postponing marriage and pregnancy, multiple sexual partners, all of which are considered to conflict with their possibility of getting pregnant. (v) Adoption is still an option to solve the problem of childlessness. (vi) Finally, masturbation was accepted. Without accepting masturbation, few diagnostic and AR procedures could be done, thus many conditions could not be explained. All contemporary ARTs are based on IVF, and they rely on masturbation or surgical interventions to obtain sperm. Hence, accepting masturbation frees the path for IVF (and therefore ICSI, PGD, MRT, etc.) to be practised.

Details of a National Portrait

I want to turn back to the national portrait of Mexico's system of assisted reproduction. Now lets focus on the logo of the Consejo Nacional de Población (National Population Council, CONAPO). What can this logo tell us about the way the ideal family composition has changed, how does it reflect and participate in the changes in Mexican society. Its first logo, dating to the 1970s, depicted a gendered family of four—a woman, a man, and two children—suggesting thus a mother, a father, a son, and a daughter. All members standing straight, side by side, with the man at the centre, taller, and between the mother and the children. After all, the father was considered the centre-stone of the family, the provider. The ideal family size was four: two parents and two children. Then, the logo transformed. Keeping the four-member family stereotype, they now use stick human figurines that do not denote gender. They too are standing, but the order is different: one adult, one child, the second adult holding the second child in its arms, its a baby. This suggests a wider gap between the two children and opens the possibility for different gender combinations of both adults and children. The family is depicted sheltered within a house, highlighting the idea that having a smaller family allows for a better lifestyle. The changes in the logo could be read almost as a matter of cause and effect, because people followed the message of the family planning campaigns, family size decreased (the first logo). The benefits of this can be seen in the second logo: this family can now afford a house to live in. Today CONAPO is addressing several issues that reflect contemporary issues: migration, violence, sexually transmitted diseases, and teenage pregnancies.

References

Allende, A. C. (2012). Telenovelas en México. Nuestras íntimas extrañas. *Comunicación y Sociedad, Nueva época* (18), 205–210.

Alvarez Lezama, F. J. (1964, Mayo–Ago). El incremento demográfico y la planeación familiar. *Estudios Sobre Esterilidad, XV*(2), 64.

Baker, J. H. (2012). *Margaret Sanger: A life of passion* (Reprint ed.). New York: Hill and Wang.

Beard, L. J. (2003). Whose life in the mirror?: Examining three Mexican telenovelas as cultural and commercial products. *Studies in Latin American Popular Culture, 22*, 73–88.

Caldwell, J. C. (2001). The globalization of fertility behavior. *Population and Development Review, 27*, 93–115.

Canto de Cetina, T. E. (1994). Programas de biología de la reproducción en México y particularmente en Yucatán. *Revista Biomed, 5*(3), 141–150.

Chavez Villasana, A. (1964, Mayo–Ago). El incremento demográfico y la planeación familiar. *Estudios Sobre Esterilidad, XV*(2), 72.

Consejo Nacional de Población. (1999). Introducción. Veinticinco años de planificación familiar en México. *La situación demográfica de México*. México: Consejo Nacional de Población.

Corona Uhink, G. (1964). Palabras del Dr. Guillermo Corona Uhink Prof. de Medicina Humanistica de la Escuela Nacional de Medicina de la U.N.A.M. *Estudios Sobre Esterilidad, 15*(2), 105–116.

de Barbieri, T. (2000). Derechos reproductivos y sexuales. Encrucijada en tiempos distintos. *Revista Mexicana de Sociología, 62*(1), 45. https://doi.org/10.2307/3541178.

de Márquez, V. B. (1984). EL proceso social en la formación de políticas/ el caso de la planificación familiar en México. *Estudios Sociológicos, 11*(2–3), 309–333.

Folch-Lyon, E., de la Macorra, L., & Bruce Schearer, S. (1981). Focus group and survey research on family planning in Mexico. *Studies in Family Planning, 12*(12), 409–432.

Franco Migues, D. (2012). Ciudadanos de ficción: Discursos y derechos ciudadanos en las telenovelas mexicanas. *El caso Alma de Hierro. Comunicación y Sociedad, Nueva época* (17), 41–71.

Gire, G. de I. en R. E. A., & Radar 4. (2015). *NIÑASy Mujeres sin Justicia. Derechos Reproductivos en México.*

González de Bustamante, C. (2009). Dependency and development: The importance of TV news in the history of Mexican television. *Galáxia* (18), 247–262.

González de Bustamante, C. (2010). 1968 olympic dreams and tlatelolco nightmares: Imagining and imaging modernity on television. *Mexican Studies/ Estudios Mexicanos, 26*(1), 1–30. https://doi.org/10.1525/msem.2010.26.1.1.

Gual-Castro, C. (2000). Programas universitarios de posgrado en biología de la reproducción. *Gaceta Médica de México, 136*(3), S75–S78.

Gutiérrez-Sánchez, S. (2000). Transición de la alta a la baja fecundidad en México. Cuadernos de investigación, Cuarta Época, No. 12. México, Universidad Autónoma del Estado de México.

Gutmann, M. (2009). Planning men out of family planning: A case study. *Sexualidad, Salud y Sociedad - Revista Latinoamericana, 1*, 104–124.

Hernández Valencia, M., & Saucedo García, R. (2018). Contribuciones del IMSS a la medicina mundial. In T. Miguel Ortega, J. de J. Arriaga Dávila, A. C. Sepúlveda Vildósola, F. A. Salamanca Gómez, & C. Martínez Castuera Gómez (Eds.), *Pasado, presente y futuro.* Instituto Mexicano del Seguro Social. IPN.

Karchmer, S. (1999). *Instituto Nacional de Perinatología 1983–1993* (Libro Negro). México: INPer.

León Olivares, F. (2001). El origen de Syntex, una enseñanza histórica en el contexto de ciencia, tecnología y sociedad. *Revista de la Sociedad de Química de México, 45*(2), 4.

Lerner, S. (1967). La investigación y la planeación demográficas en México. *Estudios Demográficos y Urbanos, 1*(1), 9. https://doi.org/10.24201/edu. v1i01.27.

Lewkowicz, E. (2014). Rebel love: Transnational teen TV vs. Mexican telenovela tradition. *Continuum: Journal of Media & Cultural Studies, 28*(2), 265–280. https://doi.org/10.1080/10304312.2013.854869.

Martínez, D. P. (1964, Mayo–Ago). El incremento demográfico y la planeación familiar. *Estudios Sobre Esterilidad, XV*(2), 101.

Martínez Manautou, J. (2013). Departamento de Investigación Científica del IMSS Su inicio en 1966. In F. A. Salamanca Gómez, J. A. González Anaya, F. Cruz Vega, & F. Aceves Ávila (Eds.), *Investigación en salud*. México, D.F.: Editorial Alfil: Gobierno de la República: IMSS: Academia Mexicana de Cirugía, A.C.: Fundación IMSS.

Martinez-Manautou, J., Cortez, V., Giner, J., Aznar, R., Casasola, J., & Rudel, H. W. (1966). Low doses of progestogen as an approach to fertility control. *Fertility and Sterility, 17*(1), 49–58. https://doi.org/10.1016/S0015-0282(16)35825-3.

Martinez-Manautou, J., Giner-Velasquez, J., Cortes-Gallegos, V., Aznar, R., Rojas, B., Guitterez-Najar, A., & Rudel, H. W. (1967). Daily progestogen for contraception: A clinical study. *British Medical Journal, 2*(5554), 730–732. https://doi.org/10.1136/bmj.2.5554.730.

Mateos Fournier, M. (1964). El incremento demográfico y la planeación familiar. Symposium. *Estudios Sobre Esterilidad, 15*(2), 60–64.

Murphy, M. (2017). *The economization of life*. Durham and London: Duke University Press.

Nagel, J. S. (1978). Mexico's population policy turnaround. *Population Bulletin* (Population Reference Bureau Inc.), *33*(5), 1.

Najam, A. (1996). A developing countries' perspective on population, environment, and development. *Population Research and Policy Review, 15*, 1–9.

Orozco Gómez, G. (2006). La telenovela en México: ¿de una expresión cultural a un simple producto para la mercadotecnia? *Comunicación y Sociedad* (0188–252X)(6), 11–35.

Ortega, M., Arriaga Dávila, J. de J., Sepúlveda Vildósola, A. C., Salamanca Gómez, F. A., & Martínez Castuera Gómez, C. (Eds.). (2018). Contribuciones del IMSS a la medicina mundial. *Pasado, presente y futuro*. Instituto Mexicano del Seguro Social. IPN. Retrieved from https://www.ipn.mx/assets/files/secgeneral/docs/SG/Libro%20IMSS%2075%20a%C3%B1os.pdf.

Paz, S. (2015, February 5). *EL mexicano que detonó la revolución sexual*. Retrieved October 29, 2018, from http://www.conacytprensa.mx/index.php/sociedad/personajes/154-el-mexicano-que-detono-la-revolucion-sexual.

Pearson, R. C. (2005). Fact or Fiction? Narrative and reality in the Mexican telenovela. *Television & New Media, 6*(4), 400–406. https://doi. org/10.1177/1527476405279863.

Pérez, Palacios y Ulloa Aguirre. (1989). In G. Soberón, J. Kumate, & J. Laguna (compliladores). Cuahutemoc Valdés (coordinador). *LA Salud en México: Testimonios 1988. Especialidades Médicas en México.*

Pick de Weiss, S. (1987). Actitudes, conocimientos y conductas de planificacion familiar en México: una década de investigación psicosocial (1975–1985). *Revista Mexicana de Psicología, 3*(2), 155–160.

Rabell, E. (1964). Palabras del profesor y licenciado Enrique Rabell del consejo nacional técnico de la educación de la Secretaría de Educación Pública. La explosión demográfica y la acción educativa del régimen. *Estudios Sobre Esterilidad, 15*(2), 77–80.

Reinhold, R. (1979, November 5). Mexico's birth rate seems off sharply. *The New York Times.* Retrieved from The New York Times Archive.

Segal, S. (1966, June). Family planning in national health programs. *Bulletin of the New York Academy of Medicine, 42*(6), 447–453.

Sierra García, F. (1964). Incremento demográfico y planeación familiar. *Estudios Sobre Esterilidad, 15*(3), 131–137.

Soto Laveaga, G. (2007). "Let's become fewer": Soap operas, contraception, and nationalizing the Mexican family in an overpopulated world. *Sexuality Research and Social Policy, 4*(3), 19–33. https://doi.org/10.1525/srsp.2007.4.3.19.

Soto Laveaga, G. (2009). *Jungle laboratories: Mexican peasants, national projects, and the making of the pill.* Durham and London: Duke University Press.

Stone, A. (1953). Problemas de Fertilidad, Esterilidad y Población. *Estudios sobre Esterilidad, IV*(2), 66.

Tate, J. (2007). The good and bad women of telenovelas: How to tell them apart using a simple maternity test. *Studies in Latin American Popular Culture, 26,* 97–111.

Tatum, H. J., & Delgado-Garcia, R. (1968). Research on physiological aspects of reproduction. *The Milbank Memorial Fund Quarterly, 46*(3), 121. https://doi.org/10.2307/3349318.

Turner, F. C. (1974). *Responsible parenthood: The politics of Mexico's new population policies.* Washington, D.C.: American Enterprise Institute for Public Policy Research.

Valdés, L. M. (1980). Ensayo sobre política de población 1970–1980: Planificación familiar. *Estudios Demográficos y Urbanos, 14*(4), 467–480. https://doi.org/10.24201/edu.v14i04.471.

Vallarta-Vázquez, M. (2005). El consentimiento informado: un derecho reproductivo en México. In M. Torres (Ed.), *Nuevas maternidades y derechos reproductivos* (pp. 239–274). México: El Colegio de México.

Welti-Chanes, C. (2011). *La demografía en México, las etapas iniciales de su evolución y sus aportaciones al desarrollo nacional, 69,* 39.

Zárate, A. (2015). Crónica acerca del Doctor Jorge Martínez Manautou, médico ilustre del IMSS. *Revista Médica del Instituto Mexicano del Seguro Social, 53*(2), 254–255.

Zárate, A., & Basurto-Acevedo, L. (2013). Notas históricas sobre la investigación científica en el IMSS. *Revista Médica del Instituto Mexicano del Seguro Social, 51*(6), 650–655.

Zárate, A., Canales, E., & MacGregor, C. (1976). *Esterilidad e Infertilidad.* México: Ediciones Científicas. La Prensa Médica Mexicana, S.A.

Zárate, A., & MacGregor, C. (Eds.). (1987). *Manejo de la Pareja Estéri. Un libro para facilitar el tratamiento de la esterilidad.* México: Editorial Trllas.

Zavala de Cosío, M. E. (1989). Fecundidad: dos momentos en la transición demográfica. *Demos, UNAM e-Journal* (2), 6–7. México, UNAM. www.ejournal.unam.mx.

Zavala de Cosío, M. E. (1992). *Cambios de fecundidad en México y políticas de población.* México: El Colegio de México/Fondo de Cultura Económica.

4

Interest in Assisting Reproduction

07:17:47 a.m. For four minutes, Mexico City's life stood still. It could not move. The earth under its feet, moved; its buildings and its houses, moved; its streets and its cars, moved; its trees and its children, moved; all moved, and moved, and moved, until it was brought down to its knees. Everyone and everything were kneeling. Somewhere praying, others were holding on because many had fallen. This is how I remember the early morning of Thursday 19 September 1985.[1] I was eleven.

Accuracy was not what dominated the aftermath of this earthquake. Nobody is sure of how many buildings and people were affected, how many were damaged beyond repair, or how many died (yes, buildings can die). Among the unrecoverable were the Gynaecology Hospital and the Scientific Research Unit of the National Medical Centre, a medical complex housing the most cutting edge medical research in the country,

---

[1]The earthquake was produced by the clashing of the Cocos and the North America plates, 18 km deep. It lasted for 4 minutes and was calculated as having a magnitude of 8.1 in the Richter scale. The material loss was estimated at 5 million dollars. An estimated 4500 hospital beds were lost when needed the most. The number of lost lives remains unknown (Cicero Sabido, Padua Gabriell, Rodríguez Martínez, Toledo, & Yáñez Villar, 1986).

© The Author(s) 2020
S. P. González-Santos, *A Portrait of Assisted Reproduction in Mexico*,
https://doi.org/10.1007/978-3-030-23041-8_4

inaugurated only 22 years before. Because these buildings endured kneeling and because its people were brave, its patients, workers, documents, and some of the equipment were safely evacuated. Nevertheless, three months after the earthquake, hundreds of kilos of dynamite imploded their remains leaving hundreds of its people in need of a new workplace, among them, two young physicians Orlando Martin Espinoza and Alberto Kably Ambe.

Martin and Kably, together with a few other embryologists and physicians,[2] were to become key figures in the establishment of the first clinics and services to offer high complexity assisted reproduction. These men are considered to be among the first to get involved in developing the knowledge and technical skills to perform IVF and to gather the material, political, and economic infrastructure to establish the first embryo labs and IVF services. This chapter is about their stories, the stories of their teams and their research, of the buildings and institutions where they worked, and the stories that made all those stories possible. It is also about the medical association I have been following, the AMEE-AMEFRH; about how it transformed its focus from thinking about the population crisis to thinking about involuntary childlessness. In other words, this chapter is about how high complexity ARTs became available in Mexico.

What exactly were high complexity ARTs at that time? They were procedures, such as GIFT (gamete intra-fallopian transfer) and IVF that involved manipulating gametes outside the body. In the case of GIFT, gametes were taken out of the male and female bodies, prepared, and then transferred back into the fallopian tubes. The idea was that the preparation done to the gametes would increase the possibility of fertilisation. In the case of IVF, gametes were also extracted from the bodies, prepared but left to fertilise outside the body. If fertilisation occurred and cell division started, the embryo(s) was (were) transferred back into the woman's body (see Appendix A for a more detailed explanation). These procedures require a team of people with diverse knowledge sets,

---

[2]For example: Alfonso Gutierrez Najar, Genaro García Villafana, Roberto Santos Halisack, Samuel Hernandez Ayup, Pedro Galache, Ricardo Asch, José Balmaceda.

working together, in spaces with specific characteristics, and with specific tools and machinery. The diversity of knowledge sets required, the tools and machineries used, and the reach of these technologies have changed since these first attempts in the eighties.

I have heard Kably speak many times since 2007, but it was not until 2016 that I had the opportunity of approaching him directly. Our first encounter took place during a closed-door meeting where physicians, groups of interest, and policy makers were preparing a document to establish the official standards of practice for assisted reproduction (the NOM on assisted reproduction).[3] After this first encounter, he has given me the opportunity of visiting his clinic many times, interviewing him for long hours, and exchanging emails. As opposed to Kably, I had never seen Martin before 2016. I met him at a special event held in honour of the two embryologists who had been responsible for the first high complexity assisted reproduction successes in Mexico: Orlando Martin and Genaro García Villafana. I managed to catch Martin as he was rapidly leaving the premises and give him a very short explanation of who I was and what I wanted, he quickly agreed to a phone interview. After our first two-hour phone interview, he agreed to a second interview in his office. Since then we have kept in contact via email and phone messages. This chapter is also about Genaro García Villafana, an embryologist working in Monterrey. I met and spoke to García Villafana at several conferences and courses I have attended along the years, however, our only proper interview has been conducted on the phone. Since then, and like with other embryologists, my contact with him has been via phone messages.

After telling their stories, I turn back to the association to see how they incorporated ARTs into their objectives. To track this I look at the books and articles they published during this period. The chapter is built by interweaving the stories presented at conferences, recounts written in articles and book chapters, news that was covered in newspapers and magazines, the memories physicians, embryologists, and administrative staff shared with me throughout several years of

---

[3]NOMs are Official Mexican Norms; these are stipulations of how procedures have to be done.

interviews, and phone messages. I am not uncovering any hidden story, I am not revealing unspoken tales, and I am not including anything they said had to remain anonymous, all I am doing is putting together pieces of a puzzle that were scattered and offering my own views about them. Therefore, I have decided to keep their real names, with their permission.

The Gynaecology Hospital and the Science Research Unit at the Centro Médico Nacional

> Latin America's most advances medical centre died after 22 years. The most important hospital complex [...] IMSS's National Medical Centre [...] This gigantic infrastructure that was able to bring together the latest advances in medicine, the most modern and sophisticated equipment, lasted 22 years. The earthquake left it standing but useless. (Salvador Corro, *Proceso*, 28 September 1985)[4]

The Centro Médico Nacional (National Medical Centre) was commonly described as the most active and productive scientific centre in the country: "from its inauguration up until the 19th of September 1985 at 13:30, when the last patients were evacuated, the Hospital never stopped working" (López-Llera, 2003) and still years after 1985, researchers kept on publishing papers on work they had done during their time at the different areas of the National Medical Centre.[5] It was in this emblematic biomedical centre where some of the first experiments with oocyte retrieval and important research in the field of the physiology of reproduction were being done. Specifically, at the gynaecology and obstetrics hospital, commonly known as Gineco 2, and at

---

[4]"A los 22 años murió el centro médico más avanzado de America Latina. El conjunto hospitalario más importante [...] el Centro Médico Nacional del IMSS [...] La gigantesca obra, que logró reunir los últimos adelantos de la medicina y los equipos más modernos y sofisticados, duró 22 años. El sismo lo dejó en pie, pero inservible".

[5]See Chapter 3 for more on the National Medical Centre.

the Scientific Research Unit (Unidad de Investigación Científica), both part of the centre but each with its unique institutional history[6] (Chavarría Olarte, Gallegos Cigarroa, & Reyes Fuentes, 2018). Since its creation in the early 1960s under the direction of Martinez Manautou, the Scientific Research Unit had been working in the field of reproduction, most of the time with financial support from the Ford Foundation (Zárate & Basurto-Acevedo, 2013). With its twists and turns depending on who was directing it, it always followed an endocrinological perspective. The Scientific Research Unit was always active in developing knowledge and training people who would later on become pioneers in the area of assisted reproduction. Among the many researchers and students that worked there were Aquiles Ayala—author of one of the first papers on IVF[7]—Samuel Hernandez Ayup—whom I will mention later on as a member of the Monterrey AR team who achieved one of the first successful pregnancies with ART in the mid-eighties, and Orlando Martin (Zárate & Basurto-Acevedo, 2013). Its last director was Arturo Zárate (1982–1988), an endocrinologist who was trained with Robert Greenblatt in the Medical College of Georgia.

The Gineco 2 was the first Mexican hospital to join the areas of gynaecology and obstetrics. Among the different specialty services it offered were the Fertility Control service, which since 1963 was supplying contraceptive methods to women for whom pregnancy meant a risk, and the Marital Sterility service, which offered infertility care. (As I discussed in the previous chapters, throughout the 1940s–1970s physicians used the terms marital sterility or conjugal infertility, these terms encompass a particular perspective of reproduction.) At the time of the earthquake, Alberto Alvarado Durán was the director of the Gineco 2 and Alberto Kably Ambe was the head of the laparoscopy service. While working

---

[6]Workers of the Gineco 2 have a strong group identity. Up until 2015, they still got together for conferences and social gatherings. There is interesting work around the institution as a healthcare facility and also as a political figure. For more on this see: http://www.gineco2.com.mx/ with its last actualization in 2015.

[7]Ayala, A. R. Montaje de tecnología para la fecundación in vitro. Memoria del curso "Avances en Biología de la Reproducción". Asociación Mexicana para el Estudio de la Fertilidad y Reproducción Humana. 1985, 253–256. Mentioned in INPer's "Black Book".

there, Kably became interested in IVF. Like the rest of the National Medical Centre, the Gineco 2 was a teaching hospital and a research centre; hence, it had access to the journals and books that otherwise would be difficult to get (we are still in a pre-internet time). This access allowed Kably to get hold of and read about the work being done by Steptoe and Edwards. "Together with some of my colleagues [Alejandro Reyes Fuentes,[8] Carlos Hinojosa Marte, and García Luna], we presented a project to learn and perfect the IVF technique. [...] We justified the importance and pertinence of the project highlighting the objectives of the Hospital: research, teaching, and caring for their patients" (Kably, Interview, 2017). Developing the IVF technique would help the hospital keep up to date, which was after all a teaching and research facility, and it would allow them to treat the growing number of patients who had infertility problems. Alvarado Durán approved the project.

The team read what the North American and British groups were publishing and based on that, they started to experiment with the oocyte retrieval and managing techniques these articles described. They worked on women who were undergoing tubal ligation as a contraceptive method. (Kably was very emphatic in clarifying that this project was done with the woman's consent and had been approved by the Hospital's authorities.) As part of the procedure, Kably would capture the oocytes and hand them over to Reyes Fuentes, who was in the lab learning to manipulate them. At this stage, they were not aiming at fertilisation, only at perfecting the retrieval and manipulation technique. The results obtained after their first year of oocyte retrieval and lab manipulation were presented at the annual meeting of the Gynaecology and Obstetrics Association in 1984. Unfortunately, because they only gave an oral presentation, never published this work, nor did they

---

[8]Alejandro Reyes Fuentes studied medicine at UNAM between 1962 and 1970, he then did a fellowship in reproductive biology at the University of Pennsylvania between 1972 and 1973. When he returned, he began working at the National Medical Centre. He is a former president of the Academy of Surgeons. His area of research was related to andrology and the morphological and functional features of the sperm.

produce a poster, there is no registry of this presentation. By June of 1985, they had enough skill to offer IVF clinically so, together with the Scientific Research Unit, where research on hormones was being carried out, they began the programme of IVF. They had started their first cycle with a cohort of patients when the earthquake hit.

During the first half of the 1980s, the National Medical Centre was developing research and clinical practice in the area of family planning and infertility care. Because this was being done in two different areas, there were two distinct perspectives being followed: the endocrinological perspective at the Scientific Research Unit and the gynaecological perspective at the Gineco 2. These two perspectives also parted in terms of their daily activity; the former oriented towards basic research, focusing on the molecular aspects of reproduction, and the later aimed at the clinical activity, caring for patients facing infertility in need of assisted reproduction (Morán Villota, Arechavaleta Velasco, & Maya Núñez, 2013).

From the National Medical Centre-Gineco 2 to INPer

> "There is very little written about the work on IVF you and your collaborators were doing at the National Medical Centre" I said to Kably during one of our long interviews. To which he answered: "True, there is little published because we were just starting, we were still trying the techniques when the earthquake happened and everything was lost, the group had to split. [...] I am the only one who continued working on assisted reproduction. The hospital was severely damaged and was demolished a couple of months after. Everything ended then and there. That is why there is so little registry, why people don't talk about it, but those were the first attempts at IVF in the country". (Kably, Interview, 2017)

Alberto Alvarado Durán and Alberto Kably Ambe arrived to INPer after the National Medical Centre was lost. Alvarado Durán became the head of the Division of Specialised Clinics (División de Clínicas Especiales). Kably wanted to continue his work on IVF. For this to be possible he had to convince two people to allow him to set up an IVF service: the head of the hospital, Roberto Karchmer, and the Health

Secretary, Jesús Kumate.[9] "I arrived to Perinato [INPer] and spoke with Dr. Karchmer, I offered him to set up the AR program. I used two main arguments: one, maintain the institute as the top institute in reproductive matters, and two, offer these services to the large number of patents that needed these services. [...] I convinced Dr. Karchmer" (Kably, Interview, 2017). Although Karchmer believed assisted reproduction would never proliferate in countries such as Mexico "because they require a lot of human and material resources" (Karchmer, 1989), he agreed that it was necessary to have these services in public hospitals since these are also research and education institutes: "Even in under-developed countries, training people in these technologies is justified, otherwise the technological gap between developed and underdeveloped nations will remain and increase [...] Given Mexico's problems of overpopulation, its priority is contraception. Nevertheless, family planning schemes, as the one implemented in Mexico, have to contemplate not only contraception but everything to help the development of the family, which includes helping couples limit their reproduction when they have reached the desired family size, and also achieve the desired family size when facing infertility" (Karchmer, 1989: 102). Then, with the INPer's approval, he went to see the Health Secretary, Jesús Kumate, who after listening to these arguments also found the project worth his support. Having Kumate's support was indispensable, since INPer is a state-run institution.

Once the programme was approved, the next step was setting up the actual service. This meant finding the funds to pay for the material infrastructure required: "...we got financial support from two places, the National Lottery, where Rogozinski, a high-school friend of mine, was director, and from Serono Laboratories (now Merck). With the money they gave us, we were able to buy an ultrasound, incubators, working benches, microscopes, all the necessary equipment for the lab. We put together an operating room in the third floor, where it still is today. You might still be able to see the plaque with the inscription: *This laboratory*

---

[9]Samuel Karchmer was a gynaecologist who focused most of his work on perinatal issues and was member of the AMEFRH. Jesús Kumate was a general surgeon and politician.

was donated by the National Lottery and Serono Laboratories. [...]" (Kably, Interview, 2017). In addition to the material infrastructure, there was the need for the human infrastructure. Two younger doctors also became part of the AR team: Paolo Di Castro Stringher, who was to work in the embryo and gamete lab, and Claudio Serviere Zaragoza. Finally, there was the need for a more solid epistemic infrastructure; if they wanted to offer a clinical service, they needed more training.

Performing assisted reproduction requires a multidisciplinary team with training in reproductive endocrinology, reproductive surgery (laparoscopy and pelvic surgery), andrology, tissue cultivation, and gamete maturation and fertilisation (both in humans and in animals) (Karchmer, 1989: 102). Karchmer and Kably invited two prominent Latin-American physicians—the Argentinian Ricardo Asch and the Chilean José Balmaceda—to work with the staff and students of the area. This would not only help train the team, it would also strengthen the area of ARTs. At that time, Asch and Balmaceda were working in Texas and had become well known in the field because of their work with GIFT. They became key figures in the process of developing the Mexican AR system because they helped train embryologists and physicians at INPer as well as at private clinics in Mexico City and Monterrey. According to several people from the field, the reason for their constant involvement with training teams in Mexico was that they were Latin American. They could communicate with the Mexican teams easily because there was no language barrier, and the cultural differences were fewer than with Anglo or with European colleagues thus, they would take into account the cultural elements shaping the patient-physician relationship, the sum of these elements meant they were able to make ARTs accessible to the local gynaecologist and embryologists (Carballo, Interview, 2017). Finally, there was a sense of camaraderie, of familiarity that did not exist with the North American physicians: "With them there were no feelings of superiority [...] it felt good to help and be helped by a fellow Latin-American" (Kably, Interview, 2017).

A third expert was brought to INPer to train the embryology staff, this was Mina Alikani. Today she is a renowned embryologist who has worked with key scientists such as Jacques Cohen (responsible for the first oocyte transfer experiments) and Alan Trounson from Monash

University (Australia). Back in the 1980s, when she came to INPer, she had just recently graduated and was beginning her professional carrier. She is one of the first women to be mentioned in these stories. It is worth noting that she was brought to work with the embryologists, an area that from the beginning was much more gender inclusive. For example, even today, female physicians in the Mexican field of assisted reproduction are underrepresented, and they have found it difficult to reached high ranks in the association. There has only been one woman president, Judith Ablanedo Aguirre who was president of the association between 1994 and 1995, and only a few have become recognised figures—for example, Mirna Echeverria and Rosario Tapia, both in the field of Andrology. Within the area of embryology, there are more women and have reached higher ranks (although within the association they are still not allowed to occupy all positions).

Once the material, human, and epistemic infrastructures were ready, they began to offer these services to the public: "We didn't do much propaganda when the service opened, we simply told the people from the infertility department that they could now send us their patients for treatment" (Kably, Interview, 2017). At that point eligibility for the programme was quite straight forward: women between 20 and 39 years of age, with no other apparent health problem, with tubal obstruction, and with no more than one living child (Karchmer, 1989: 103). For the INPer directives to be able to justify the cost of the AR service, it needed to prove it was achieving something. In 1992, during the annual meeting of the COMEGO they presented their first results (published also in the journal Ginecología y Obstetricia de México). Between 1989 and 1992, they had cared for 436 couples, performed 1627 procedures, and achieved a total of 212 pregnancies via AI, 32 with GIFT (out of 175 couples who were being treated with GIFT) and 7 with IVF (out of 261 couples being treated with IVF). According to Kably "[…] these results are, beyond doubt, the keystone for the first institutional program of this type" (Kably et al., 1992).

One of the arguments to justify the establishment of AR service at INPer had been training future generation of physicians. Between 1986 and 1994, 39 people graduated from the speciality in biology of human reproduction (only 6 were women). As part of the academic activities, there were annual conferences, as part of these there was always something on assisted reproduction. Then, as part of the fifth annual meeting of InPer, Kably and Delgado Urdapilleta organised a full course on the "Advances in Human Reproduction" and published a collection of the papers presented. There were papers on endometriosis, artificial insemination, sperm capacitation, laparoscopy, reproductive surgery, ovarian stimulation, and IVF. Kably authored the one on IVF. Between 1989 and 1993, the AR service and achieved 40 births using GIFT and 9 with IVF (Karchmer, 1999).

In 1994, a new administration was appointed to InPer pushing Karchmer and the team at the assisted reproduction service out of InPer. Karchmer and Kably establish a private clinic in a wealthy borough of Mexico City called Palmas. As head of the new IVF and gamete labs, they invited Esperanza Carballo, a young and energetic veterinarian who had worked in the area of reproduction at a dairy farm, thus had experience managing gametes and embryos. When this dairy farm closed, she went to InPer to work with Carlos Villanueva in the area of reproductive physiology. In 1998, the team moved again, now to the new Angeles Hospital located in Interlomas, where they have been since.

From the Scientific Research Unit of the National Medical to the Humana Hospital

After the earthquake, Martin looked for work options, but none of the ones available suit his research interest. Then, he remembered he had heard that Alfonso Gutierrez Najar and Roberto Lichtenberg—physicians whom he previously knew from National Medical Centre—were looking for someone to help them with the work they were doing on reproduction at the newly opened hospital Humana, so he asked them for an appointment. After two months of calls and arrangements with Marta García, the clinic's secretary, he finally got a chance to speak to Gutierrez Najar. "What do you do?" he asked. Martin told him his story: he had graduated from medicine in 1976 (UNAM), then had obtained a Ford Scholarship to go, as a research fellow for a year, to the Washington University Medical School in San Luis Missouri (USA).

There he developed a project focused on the different factors involved in ovulation, particularly the effects of steroid hormones in the development of the plasminogen. Upon his return, he got a job at the Scientific Research Unit of the National Medical Centre, with Zárate, where he worked for the following years researching hormonal mechanisms in reproduction. When Gutierrez Najar heard his story, he exclaimed: "*Me caes como anillo al dedo*,[10] I need a lab expert with your experience more than a clinician. Someone just like you. You start tomorrow!".

Alfonso Gutierrez Najar, an energetic and charismatic man, studied medicine at the UNAM in the fifties. During his school years, he met various important national and international figures in the field of endocrinological gynaecology which influenced him profoundly, igniting his interest in the function of hormones in the reproductive process. Among these was Boris Rubio Lotvin, a Mexican gynaecologist working on the endocrinology of reproduction, member of the AMEE, and known for his research and clinical practice focused on contraceptive methods. Then, he did an internship at Johns Hopkins with Emil Novak where he met Georgeanna Seegar Jones. Upon his return to Mexico in 1963, he found a job at the Hospital de la Mujer, of which he soon became director. While there he met Martinez Manautou, who at that time was working at Syntex. Together they began to research the use of hormones to control reproduction, a project financed with money from the Population Council and the Ford Foundation. As part of this project, Gutierrez Najar opened a fertility clinic at the Hospital de la Mujer where he offered both family planning options and infertility treatments (mostly artificial insemination) (Lerner, 1967; Martinez-Manautou et al., 1967; Reinehart, 1977). All was well until President Luis Echeverria took office in 1970. Faithful to his belief that "to govern is to populate", he installed very strict pronatalist policies and shut down all public programmes related to family planning, including the fertility centre at the Hospital de la Mujer (Acevedo, 2003). Echeverria

---

[10]The Mexican colloquial expression meaning "you were just what I was looking for" or "you fit perfectly".

never believed that, in addition to offering contraception, Gutierrez Najar was helping people conceive: "he said that all we were doing was contraception disguised as fertility programs" (Gutierrez Najar, Interview, 2007). He was forced to leave the hospital. Despite these setbacks, his work in family planning gained him international recognition[11] which led him to become a consultant for the World Health Organisation. During these activities, he met Patrick Steptoe, with whom he kept in contact for many years and who planted in him the curiosity for IVF.

Gutierrez Najar lived and spoke of contraception and reproduction as two sides of the same coin, he would say that knowledge on the physiology of hormones within reproduction would help either halt or promote it, depending on what was desired. After being mostly engaged with one side of the coin for nearly two decades, in the eighties he flipped the coin and become mostly engaged with assisted reproduction. Since the early 1980s, he was treating women who were having problems conceiving with clomiphene citrate. Seeing that this problem was frequent enough and given the work done by Steptoe and Edwards, he thought that it would be a good idea to establish a fertility clinic. He went to the USA, together with his son Javier (also a physician and with whom he was going to set up the clinic), to take a course in microsurgery, which are skills needed for egg retrieval and embryo transfer. Once back, he approached the American British Cowdray Hospital (ABC Hospital)[12] with a project to open a fertility clinic, but one member of the board did not approve of these practices, so his proposal was rejected. Soon after he was approached by people who were recruiting

---

[11]He is known internationally for his work in surgical tubal sterilisation, particularly the culdoscopic technique (a.k.a culdoscopic tubal clipping) for which he developed a special forceps that bears his name: the Gutierrez Najar forceps (Frusch, Interview, 2018). This technique, which requited less time, local anaesthesia, and shorter recovery period, was used in many countries where family planning campaigns were implemented.

[12]The ABC Hospital (American British Cowdray Hospital) is a private hospital set up in 1952 by the British and American community. A few years later Alberto Kably Ambe also approached the ABC Hospital with the idea of a fertility clinic, and again they refused. It was not until the early 2000s when Carlos Navarro brought them the idea again that they agreed to opening a clinic as part of a modernising project that included opening a new hospital.

renowned physicians from several specialities (urology, cardiology, oncology, orthopaedics, radiology, plastic surgery, and gynaecology[13]) to work at their new hospital: Humana. Gutierrez Najar was offered to be the head of the fertility clinic, an offer he accepted.

Humana was a high-end, for-profit, US hospital corporation.[14] In 1979, they decided to open a branch in Mexico City thinking it would be a profitable project. Humana marketed itself as a hospital that brought together, on the one hand, cutting edge equipment and the best USA- and Europe-trained Mexican doctors, an on the other, the service of a five-star hotel. For example, they offered private rooms with television, a French wine list, bilingual secretary services, and a heliport. Their target was the wealthy sector of the Mexican population who commonly sought medical treatment in the USA. This new hospital would have 200 beds and would be located in the south part of the city, in a 47,000 sqm plot of land that used to be a farm, next to a river, in an area called El Pedregal[15] (Ortiz Pinchetti, 1984; Stomer, Girault, & Wilson, 1984; Vera, 1986).

Construction began late in 1981 and by 1984 it was ready to open to the public (Stomer et al., 1984). The context of the hospital's inauguration was complex. The corporation boosted that they were among the best healthcare services worldwide, with 85 hospitals in the USA and 5 in Europe. However, in a country facing a great financial crisis, the idea of having a foreign for-profit hospital angered many within the medical community (Ortiz Pinchetti, 1984).[16] Some saw this as an example of the privatisation of healthcare services Mexico's health system was

---

[13]Samuel García Peláez was designated as medical director.

[14]Humana Inc. started out as a nursing home in the 1960s. In 1968, they bought their first hospital, by 1974 they had turned into a hospital owning business and changed their name from Extendicare to Humana.

[15]El Pedregal—area of rocks—is the name of this area of the city in reference to its volcanic rock resulting from the eruption of the Xitle volcano some 25,000 years ago.

[16]The eighties brought more than just a geological earthquake that collapsed the city, it also brought an economic quake that collapsed the economy. After twenty-five years of relative political stability, economic growth, and investment in infrastructure, the "Mexican Miracle" was no more. Between 1976 and 1982, Mexico's peso faced two plummeting devaluations: the first in 1976, triggering capital flight and doubling Mexico's debt, and the second in 1982, when the banks were nationalised.

moving towards. They feared that they were there only to take advantage of the cheap labour offered in Mexico. Others saw the situation as immoral and unfair given that the physicians who would work there, charging high prices that the majority of the population could not afford, had studied at the National University with public funds.[17] There were also criticism and worries concerning the potential users. Some physicians feared that there would be people who would spend large sums of money and acquire un-payable debt just to receive (what was being promoted as) "the best treatment" brought from the USA. Furthermore, unless they made it also a teaching hospital, Humana was not going to contribute to scientific research in Mexico, as they claimed. Others were more cautious and reminded critics that Humana was a private investment, it was not done with public money, so it did not cost the people; that it was going to be targeted to the wealthiest who, instead of going abroad in search of health care, would use the services here, giving jobs to local people; and that, since the thirties, the Mexican healthcare system has been a mix of public and private services.[18] In spite of these criticisms, Humana inaugurated in April 1984 and Gutierrez Najar started working at the fertility clinic together with Roberto Von Lichtenberg, Dutch physician and midwife Anna Frusch, and geneticist Mabel Cerrillo-Hinojosa. They were only missing someone with the necessary experience to take charge of the laboratory. This person turned out to be Orlando Martin.

---

[17]Mexico has a tuition-free education system. Public schools and universities receive tax money from the government. Regardless of the student's socioeconomic level or income, those who study at the public schools or universities do so tuition-free. This includes UNAM's School of Medicine.

[18]Since the thirties there had been privately owned hospitals for specific communities such as The Spanish Hospital (1932) and the ABC Hospital (1952), neither of these however were for-profit. According to Quijano, a physician quoted in an article in Proceso, there had been two prior attempts at establishing a high-end private healthcare service, but those had been a Mexican initiative with Mexican investors. The first was in the early 1970s (1973–1974), it was organised by Agustin Legorreta with financial backing from the Banco Nacional de Mexico and with cardiologist Clemente Robles as the medical leader. This project never lifted because some members of the bank did not support it and then the 1976 crisis hit. The second effort was Manuel Campuzano's project of Medica Sur, which eventually was established and is still offering service today.

Upon joining Gutierrez Najar's team, Martin's first task was to make a list of all the things he needed for the lab. "Within a week I started receiving the materials. Humana bought everything; the majority of the things came from abroad. The equipment took a little less than four months to arrive. There were some parts of the equipment that were damaged, and other things were detained at customs, but by the end of the year we were practically set" (Martin, Interview, 2018). Looking back, Anna Frusch said: "that first lab was quite rudimentary, it was a beautiful lab, set up with great care and dedication, but at that time nobody was talking about the good laboratory practices nor about the guidelines regarding the management and control of light, air, and temperature, layout and architectural finishings. [...] We did things like turning around the ultrasound screen to see the image in a way that was better for the procedure" (Frusch, Interview, 2018). In the second half of the eighties, the embryo labs were nothing like they are today. Back then there were not as many ready-made off the shelf tools for IVF and assisted reproduction in general, embryologists needed to be creative and find ways to solve technical problems (Carballo, Interview, 2019). Martin was sent to the London branch of Humana, the Wellington Humana Hospital, to work with Ian Craft and thus "acquire experience" in the different aspects of the ARTs. "I saw how they did the GIFT, how they prepared the sperm samples, how they captured and prepared the eggs, how they prepared the medium; because back then, we had to prepare the medium at the clinic, it was not bought readymade as we now do" (Martin, Interview, 2018). In order to be able to prepare usable medium, they needed to secure purified water, something which turned out to be difficult problem to solve. When he came back, he trained two more embryologists: Lorena Diaz and Gaby Aguilar.

A few years after the hospital inaugurated, in the middle of Mexico's major financial crisis and changes in the US taxation laws for foreign investment, Humana was put up for sale in a public auction. According to the hospital's financial director (J. Greg Britt), they were losing money from the very onset of the project due to the magnitude of the country's crisis (Vera, 1986). Others say it was a financial move. Some thought that the problem went beyond the economic crisis. When the hospital went on sale, the representative of the hospital's medical community, Federico Ortiz Quezada, spoke openly about why he considered Humana to be doomed from the beginning: "They came with an absurd image, wanting to sell a health service as if it were a jewel or a perfume. They were ill-advised. They did not see that a USA style medicine could not be functional in our country" (Ortiz Pinchetti, 1984; Vera, 1986). Identifying different styles in medicine is something that is constantly mentioned by the community of AR practitioners. When pointing this out, practitioners are commonly making reference to the healthcare systems in the different countries, and how these articulate with the AR services, to the regulatory and legal infrastructure surrounding (or not) the AR services, and to the economic and political environment that also affects the practice of the AR services. This particular comment is making reference to the state of the medical practice, which at that moment, had still not been commodified to the extent it is today; health services did not have to be luxurious, they have to be good. For these or other reasons, Humana Inc. sold two of its branches, the one in Mexico and the one in the UK. Apparently, at first nobody wanted to buy the Mexican branch, not even the government. Then, a group of doctors who worked at the hospital (among them was Gutierrez Najar) together with some investors placed an offer, but the final and higher bid came on 10 December 1986 from the Mexican wealthy entrepreneur Olegario Vazquez Raña. Buying Humana was the first move towards what ultimately became his large network of hospitals, the Grupo Angeles.[19]

Gutierrez Najar offered the new owners to buy the fertility clinic in monthly payments during the following six years. Anna Frusch remembers: "They sold us the equipment as if it were new, when it was in fact two years old, and they sold it in dollars. Imagine this in

---

[19]The family Vázquez Raña is one of the richest and most powerful families. They own (or have owned) newspapers, radio stations, television channels, hotel chains, banks, insurance companies. They have invested in the country's infrastructure (airports and public transport) and have been in very close relationship with the PRI, the party that governed the country during the twentieth century. In 2018, the group held a total of twenty-four hospitals, ten in Mexico City and fourteen in the rest of the country. Grupo Angeles was the first network of private hospitals to exist in Mexico.

the middle of the crisis!" Putting together a lab is expensive, and having it running is equally so. "The first year we had ten patients. This made sustaining the clinic very difficult. We had to pay salaries, the equipment, materials, plus the money we owed to Vazquez Raña. All this with only ten patients. So Alfonso [Gutierrez Najar] and Roberto [Lichtenberg] had to work very hard in their other areas [gynaecology patients not in need of ARTs] to make ends meet. Then, by the time we finished paying the money we owed the Vazquez Raña, the equipment was to old! things change very quickly in the field of assisted reproduction. I know people thought otherwise, but we never made millions, people thought we were rich because Alfonso always had patients and worked very much. But the truth is the clinic was never a sound economic business" (Field notes, interview with Frusch, 2018).

It was in this context that they achieved their first success: Andrea, conceived in 1987 through GIFT, was born on 23 March 1988.[20] Following Frusch, I too want to stress that these achievements were not accomplished individually but through constant teamwork. When Frusch was telling me the story, she emphasised it was *they* who achieved the birth, *they* to her were the three men: "Alfonso was not the first at IVF, it was Alfonso, Orlando, and Roberto, the three of them" (Frusch, Interview, 2018). I would add the women working there: Frusch, the other two embryologists Lorena and Gaby, and the nurses. Female actors tend to be omitted, even by the women themselves.

Gutierrez Najar's clinic also became a teaching clinic, working in conjunction with the medical school at UNAM and at La Salle. Many of the first generation of specialists now working in Mexico City and across the country were trained there. In 1999, Martin left the clinic and went to establish IVF labs and train embryologists in other parts of the country like Guadalajara (Centro de Reproducción Asistida de Occidente, with Juan Gutierrez Hermosillo), San Luís Potosí, and Sonora.

---

[20]Her two siblings were also GIFT babies.

And in the North, the IECH

The Mexican northern border covers 3145 km from the Pacific Ocean to the Gulf of Mexico and has six states: Baja California, Sonora, Chihuahua, Coahuila, Nuevo León, and Tamaulipas. Samuel Hernández Ayup was born in Tijuana (Baja California), Pedro Galache Vega in Chihuahua (Chihuahua), and Roberto Santos Haliscak in Monterrey (Nuevo León); the three of them, together with Manuel García Martínez, Ricardo Loret de Mola Gutierrez, and Fidel Morales Caballero, created the Instituto para el Estudio de la Reproducción Humana at the Centro de Ginecología y Obstetricia de Monterrey (CEGOMSA now Ginequito) circa 1984.[21] Like INPer, since 1995 IECH has also been a centre where many AR specialists, both physicians and embryologists, go for training.

Hernández Ayup was the senior. He had studied medicine at the UNAM in Mexico City, worked at the Gynaecology Hospital 1 of IMSS where Castelazo Ayala was director and where he met Arturo Zárate. He then worked in the Scientific Research Unit of the National Medical Centre with Zárate in a project on clomiphene as an ovulation inductor. By the mid-seventies, he went to the USA to do a subspecialty in reproductive gynaecology and endocrinology with Greenblatt, Zárate's mentor. He stayed there for four years, during this time he met Ricardo Asch with whom he keep in contact for many years after. Eventually, he sent three of his junior physicians, Pedro Galache, Santos Haliscak, and García Martínez, to study with him (and Balmaceda) and acquire the skills needed for GIFT and IVF.

When they began experimenting with ARTs in the mid-1980s, Hernández Ayup and Galache Vega were performing the laparoscopy, the surgical procedures of retrieving ova, while Santos Haliscak and García Martínez were performing most of the laboratory activities (egg and sperm preparation); but neither of them were biologists or embryologists. After a year of unsuccessful attempts, they began looking for

---

[21]Later on, there was a slight change in the name of the centre, interchanging reproduction for conception, now the IECH.

someone that would take over the work done in the laboratory, someone interested in developing and working with techniques to assist fertilisation and reproduction.

It was 1986 and García Villafana was a young student, still undertaking his undergraduate degree as a clinical lab biologist. He was looking for a laboratory that would allow him to do his placement when he
heard that there was a team of physicians looking for someone to take
charge of their laboratory: "I meet them because a chemist who was
working there told me that they were looking for someone interested
in their 'test-tube baby' project, because that is how we called them
back then, test-tube babies, now we call them babies from reproduction" (Interview, 2018). García Villafana went to meet the doctors and
got the job, this is how a successful working relationship that, to great
admiration of the entire Mexican assisted reproduction community, has
lasted for over thirty years.

Like Martin, his first task was to organise the lab. While doing so,
he also learned from Santos Haliscak and García Martínez the art of
ART. Within two years, the team achieved its first successful pregnancy:
Carlos Esteban, conceived using GIFT and donated ova, was born on 24
February 1988. A year after this (and a few subsequent successful conceptions and pregnancies) García Villafana was sent to Colombia: "[…]
to Bogotá with Dr. Elkin Lucena who had achieved the first IVF baby in
Latin America[22]" (Interview, 2018). There he acquired the skills to take
on the full task of becoming the head of the embryo and gamete lab at
the IECH. (It is worth mentioning that embryologists have two major
tasks when working at IVF labs: one, the administration of the lab and
management of the equipment, and two, performing the different procedures; both things play an important part in achieving success.)

Monterrey is the capital of the northern border state of Nuevo
León, it is located in a valley at the foot of the famous Cerro de la
Silla and the Sierra Madre Oriental. It is one of the three wealthiest
cities of the country with a very close-knit conservative society that

---

[22]Conceived at the Clinica del Country de Bogotá and born on 10 January 1985 by C-Section.
The physicians Oscar Lombana, and Luz Helena Pérez also participated.

holds great pride of its city, state, and local culture. Important companies were born in Monterrey, companies that today have a national and transnational presence.[23] It is home to two important universities, one private, the Tecnológico de Monterrey (TEC) and the public the Universidad Autónoma de Nuevo León (UANL) which has one of the few University owned Hospitals.[24] Being only six-hour straight drive from San Antonio (TX), three from McAllen (TX), and only two hours from Laredo (TX), it has close contact with the southern border of the USA. It is not uncommon for those who can, to go to "the other side" (be it Mexico or the USA) to buy goods and seek services found cheaper or better than in their own city. This constant and mundane contact has made its local culture seem considerably "americanised" to the people in the centre and south part of the country.

When it came to assisted reproduction, its conservative atmosphere outweighs its "americanised" culture. The pioneering work being done at IECH in the mid-1980s was received with moral criticism. During a short conversation I had with Santos Haliscak a few years ago, he commented how he felt during those early days "When I went to Church, I felt judged and somehow attacked by the people. Then, when we started freezing embryos, some people picketed at our door and some patients told me they were against what we were doing".

The Stories of the First

During my field work, I frequently heard comments that pointed towards a dispute over who had achieved the first high complexity ART birth. People would say: "hopefully you will be able to solve the controversy" or "there are disputes over who was first". However, when I had

---

[23]Companies founded by families from Monterrey include: Cervecería Cuauhtémoc Moctezuma, a brewery founded in 1890 that today is part of Heineken; FEMSA, Coca-Cola's largest bottling company established in 1890; VITRO, a glass company founded in 1909 as the main supplier of the brewery Cuauhtémoc Moctezuma.

[24]This hospital also has an IVF clinic called CeUMER, Centro Universitario de Medicina Reproductiva de la Universidad Autónoma de Nuevo León.

spoken to those directly involved (Gutierrez Najar, Orlando Martin, Genaro García Villafana, Roberto Santos Halisack), they never seemed to imply there was a controversy. Those who spoke of a controversy, of a dispute, were people from the newer generations of physicians, embryologists, and staff. Why? This is what I believed happened.

When "the first" babies conceived using high complexity ARTs were born in Mexico, there was very little media coverage, unlike the case with Louis Brown (Challoner, 1999). Even the scientific coverage was meagre. The IECH published the paper reporting their success a year later in the national journal of gynaecology and obstetrics (Hernández Ayup et al., 1989) and the AGN only gave a presentation on the success rates at their clinic, but that was until 1992 (Gutierrez Najar et al., 1992 (6)S1: 100).

Years later, as the story of the field began to be written and these stories began to appear in academic papers and books, in the media, in interviews published the national press, and in the clinic's websites, versions, and variations emerged. For example, the book *Medicina Reproductiva* (published first in 1999 and then in 2003) presented an ambiguous narrative. In the chapter dedicated to GIFT, written by Carlos Felix Arce and Manuel García Ramírez, there was no mention of these "first babies", while in the chapter on IVF, written by the IECH team, there were a few lines on their first successful cases, but they said it happened in 1986 and they describe the technique simply as "assisted reproduction" (Galache Vega et al., 1999: 282; 2003: 364). Not mentioning the successful GIFT cases in the GIFT chapter, yet presenting one undefined success in the IVF chapter gives room for the interpretation that the first success was done with IVF. Another example of an ambiguous narrative is an article published in 2003 by the newspaper *La Jornada*. The piece was on the 15th birthday of the first "test tube baby" to have been conceived in Mexico City, with the help of Orlando Martin and Gutierrez Najar. This baby was conceived using GIFT, thus using the term "test tube baby" might be confusing. While the term dates back to a pre-IVF period, it was a term that by 1978 became strongly associated with IVF, particularly with the out-of-body-fertilisation aspect of IVF which is precisely what distinguishes these two techniques (in GIFT fertilisation takes place inside the woman's body).[25] Then, if we turn to the clinic's web-sites, the stories there told state that IECH's first successful case was Carlos Esteban, born on 24 February 1988, conceived using GIFT with donated ova; Gutierrez Najar's clinic website says that their first was Andrea, born on 23 March 1988, conceived using GIFT, and that their first successful IVF was not until 1991. A fourth example of how these stories presented variations happened during the annual meeting of the Mexican Association of Reproductive Medicine in 2017. At this event, the biologists that achieved these first successes were honoured. Sandra Cubillos, an embryologist working at a clinic in Mexico City, gave a recapitulation of the past thirty years focusing on the work done by the embryologists. During this talk she pointed out the lack of publications concerning these first successes and added "what I was able to find was that the first GIFT baby was achieved in 1987 at Gutierrez Najar's clinic, but it was not reported. The one he did report was his first IVF, which was born on the 23 of March 1988" (Field notes, AMMR anniversary event, Cubillos, 2017).

Pointing out these variations do not have the purpose of "finding out the truth" or of evaluating their accuracy. Highlighting these variations has the purpose of taking them seriously, of seeing them as part of the history of the field, and considering the context in which they were produced. Between 1986 and 1988, these were the only clinics doing high complexity ARTs. However, given that they were 900 km apart, they were not really competing for clients or reputation. As long as both received credit for being the first clinics to open in Mexico, to have trained the new generation of physicians, and for having both achieved success within the same period of time, the detail of who was before whom has been, to an extent, irrelevant.

---

[25]The term "test tube babies" appears as early as 1948 in the B movie *Test Tube Babies* directed by W. Merle Connell and produced by George Weiss. Here, the term "test tube babies" is making reference to babies conceived using artificial insemination. One of the posters promoting the movie had an illustration of a large test tube with a baby in it. Then, in 1969, the term is re-signified to make references to IVF children. For example, on the cover of the magazine *LIFE* there is reference to test tube babies now thinking about IVF.

These variations also help us think about how "being the first" is performed in these cases because, within all these different versions of the story, there are two things that never changed: first, positioning both the IECH and Gutierrez Najar's clinic as the first two private clinics to be established and as the first clinics to achieve a live birth after manipulating the gametes outside the body and second, omitting certain people and places. For example, the work that was being done in the National Medical Centre before the earthquake of 1985 by people like Alberto Kably and Alejandro Reyes Fuentes or the work being done at the National Institute of Nutrition. Because their work did not result in "a first", their contribution to the story of assisted reproduction in Mexico is commonly left untold.[26]

These narratives also show the importance GIFT, Ricardo Asch, and José Balmaceda, the physicians credited with perfecting and popularising it, had in Mexico's assisted reproduction story. During the 1980s, GIFT was gaining attention because clinics were reporting better results with GIFT than with IVF. Some considered that this higher success rates "could be related to the fact that the technique mimics the physiological steps required for human fertilization" (Borrero, Ord, Balmaceda, Rojas, & Asch, 1988; see also Wang & Sauer, 2006; Yovich, 1994). Furthermore, it was a procedure that avoided some of the moral and practical issues that emerge with IVF. For example, because fecundation takes place inside the woman's body as opposed to IVF where it takes place in vitro, it did not produce out-of-body embryos (pre-embryos), thus avoiding the conundrum of whether to dispose, donate, or freeze the ones that were not transferred. However, by the mid-1990s the use of GIFT had decreased, they were using it in only 20% of the cases. This decline could be due to two factors: one, the preference physicians had of being able to corroborate fertilisation, which they could do with IVF and two, the 1995 fertility-scandal that surrounded Asch and Balmaceda and that tainted the technique. In 1995, while working at a clinic in Irvin (California, USA), Asch and Balmaceda were accused of swapping eggs between patients in order to boost clinic's success rates (Wexler, 1995).

---

[26]In some interviews with younger physicians and embryologists, I was also told that there was other physician who claimed to have "achieved the first test tube baby". Other than these oral communications, I have found no other reference to this incident.

One of the premises this book is built upon is the acceptance of the origin story the AMMR tells about itself. In other words, that today's professional association (AMMR) is the same association that signed its constituting legal document 70 years ago, regardless of the changes it has undergone in terms of its name, location, members, goals, and even moral make-up. This book is also set upon the acceptance of the claim that there is a thing called infertility (or sterility, or sub-fertility). Finally, this book is an attempt to understand what ARTs have been in Mexico. However, following the association, the notion of infertility, and the ARTs in Mexico through time and space, it becomes clear that, like the atherosclerosis that Annemarie Mol (2002) followed at a Dutch University Hospital and the erectile dysfunction that Emily Wentzell (2013) followed at a hospital in Cuernavaca (Mexico), none of these are simple, immutable, coherent, or consistent entities. Upon close and careful interaction with them, they reveal a multiplicity, mutability, complexity, internally contradictory dynamic relational way of being in the world.

One of the ways in which the association has been able to sustain itself as one while at the same time being multiple across time and space, enduring the epistemological and material changes it has undergone, has been through the way they perform their history and their identity. This way has been a sort of selective memory-making that allows for a larger history to be told. For example, by claiming an institutional age that goes back to 1949, by mostly keeping the original logo, and by displaying the past presidents' portraits in the conference hall at AMMR's headquarters. However, they also have changed their name to highlight their new goals and perspectives, and they have edited their story of origin to omit the elements that would tension the possibility of being one and the same association (for example AMEE's rejection of AI).

Detail of a National Portrait

The detail I want to focus on now is the only picture of "the first" that has circulated in the media. This picture is of the baby conceived at the IECH using GIFT. There are five people in the picture: Pedro Galache Vega who is looking into the camera, Manuel García looking over Samuel Hernandez Ayup shoulder to see the baby he is holding, a baby that is still naked baby wrapped in a pink blanket, and Roberto Santos Halisack smiling to the camera. They are all dressed in blue-green scrubs, standing next to a crib-like table where another baby is lying, wrapped in a yellow blanket, and facing away from the camera. The setting suggests this picture was taken shortly after the birth had occurred. The other actors who were crucial for this moment to happen—the ova-donor, the mother, the father, the embryologists, and nurses—are not captured in this picture. This picture could be of any other birth. The aesthetic qualities of the image have nothing spectacular, nothing unique. This picture does not work as a proof of concept in the same sense as the photographs taken of Louise Brown, Dolly the Sheep, or even Dr. Zhang and baby Hassan. It does not work as such because IVF and GIFT had already been proven to work. These were no longer procedures under experimentation or undergoing a test. These were proven techniques. What this picture proves its that these particular physicians, from this particular clinic, in this specific city and country were able to successfully carry out a GIFT. In other words, this picture works as a proof of the team, not of the technique.

References

Acevedo, E. (2003, Marzo). Alfonso Gutiérrez Najara: Ayudando a la vida a dar vida. *Medicos de México*. Año 1. No. 10. 12.

Borrero, C., Ord, T., Balmaceda, J. P., Rojas, F. J., & Asch, R. H. (1988). The gift experience: An evaluation of the outcome of 115 cases. *Human Reproduction, 3*(2), 227–230. https://doi.org/10.1093/oxfordjournals.humrep.a136682.

Challoner, J. (1999). *The baby makers: The history of artificial conception.* London: Channel 4 Books.

Chavarría Olarte, M. A., Gallegos Cigarroa, A., & Reyes Fuentes, A. (2018). Unidad de Investigación Médica en Medicina Reproductiva. In T. Miguel Ortega, J. de Jesús Arriaga Dávila, A. C. Sepúlveda Vildósola, F. A. Salamanca Gómez, & C. Martínez Castuera Gómez (Eds.), *Contribuciones del IMSS a la medicina mundial. Pasado, presente y futuro. Instituto Mexicano del Seguro Social.* México: IPN. Retrieved from https://www.ipn.mx/assets/files/secgeneral/docs/SG/Libro%20IMSS%2075%20a%C3%B1os.pdf.

Cícero Sabido, R., Padua Gabriell, A., Rodriguez Martinez, H., Toledo, B., & Yáñez Villar, A. (1986). Efectos del terremoto del 19 de septiembre de 1985 en el Hospital General de la Ciudad de México. Algunas consideraciones. *Salud Públia México, 28*, 521–526.

Corro, S. (1985, September 29). A los 22 años murio el Centro Médico más avanzado de América Latina. *Proceso.* Retrieved from http://www.proceso. com.mx/142125/a-los-22-anos-murio-el-centro-medico-mas-avanzado-de-america-latina.

Galache Vega, P., Hernandez Ayup, S., Arenas Santillán Gómez, A., Ruy Sánchez Aguilar, M., & Santos Haliscak, R. (1999). Fertilización in Vitro. In E. Vázquez Benítez (Ed.), *Medicina Reproductiva* (1st ed., p. 279, Chap. 54). México: JGH Editores. ISBN 9706810250.

Galache Vega, P., Santos Haliscak, R., Batiza, R., Saucedo de la Llata, E., Arenas Montezco, L., García Villafana, G., & Ayup Hernández, S. (2003). Fertilización in Vitro. In E. Vázquez Benítez (Ed.), *Medicina Reproductiva* (2º). México, D.F., México: Editorial Manual Moderno, AMMR.

Gutierrez Najar, A., Navarro, M. C., Martin, O. E., & Diaz, M. (1992). Fertilización in vitro-GIFT: evaluación de una clínica 1988–1992. *Ginecología y Obstetricia de México, 60*(S1), 100.

Hernández Ayup, S., Santos Haliscak, R., García Martínez, M., Morales Caballero, F., Loret de Mola Gutiérrez, R., & Galache Vega, P. (1989). Gift: Reproductive reality for the sterile couple. *Ginecología Y Obstetricia De México, 57*, 315–319.

Kably Ambe, A., et al. (1992). *Ginecología y Obstetrica de México, 60*(Suppl 1), 93.

Karchmer, S. (1989). *Temas Selectos en Repoducción Humana Parte I.* Mexico: Instituto Nacional de Perinatología.

Karchmer, S. (1999). *Instituto Nacional de Perinatología 1983–1993* (Libro Negro). México: INPer.

Lerner, S. (1967). La investigación y la planeación demográficas en México. *Estudios Demográficos y Urbanos, 1*(1), 9. https://doi.org/10.24201/edu. v1i01.27.

López-Llera, M. (2003, February 27). *Pasajes inéditos en la vida y muerte del hospital de gineco-obstetricia del Centro Médico Nacional.* Retrieved from http://webcache.googleusercontent.com/search?q=cache:vGRZh21DxZ8J: gineco2.com.mx/fotos_archivos/Pasajeshgo2.doc+&cd=1&hl=en&ct= clnk&gl=mx&client=firefox-b-d.

Martinez-Manautou, J., Giner-Velasquez, J., Cortes-Gallegos, V., Aznar, R., Rojas, B., Guitterez-Najar, A., & Rudel, H. W. (1967). Daily progestogen for contraception: A clinical study. *British Medical Journal, 2*(5554), 730–732. https://doi.org/10.1136/bmj.2.5554.730.

Mol, A. (2002). *The body multiple: Ontology in medical practice.* Durham: Duke University Press.

Morán Villota, C., Arechavaleta Velasco, F. J., & Maya Núñez, G. (2013). Biología de la reproducción. In F. A. Salamanca Gómez, J. A. González Anaya, F. Cruz Vega, & F. Aceves Ávila (Eds.), *Investigación en salud.* México, D.F.: Editorial Alfil: Gobierno de la República: IMSS: Academia Mexicana de Cirugía, A.C.: Fundación IMSS.

Ortiz Pinchetti, F. (1984, Marzo 17). Controversia entre médicos. *Proceso* (385). https://www.proceso.com.mx/.

Reinehart, W. (1977). *La minipíldora. Una alternativa limitada para ciertas mujeres* (Anticonceptivos Orales No. 3) (pp. 61–79). Universidad de George Washington. Retrieved from https://www.k4health.org/sites/default/files/751074SPA.PDF.

Stomer, M. R., Girault, P., & Wilson, L. E. (1984). *Combined Earth and Rock Bearing Foundation—Hospital Humana Mexico City D.F.* International Conference on Case Histories in Geotechnical Engineering. 7. Retrieved form http://scholarsmine.mst.edu/icchge/1icchge/1icchge-theme1/7.

Vera. (1986, December 13). El hospital, en venta, pero nadie compra. *Proceso.* Retrieved from https://www.proceso.com.mx/145031/el-hospital-en-venta-pero-nadie-compra.

Wang, J., & Sauer, M. V. (2006). In vitro fertilization (IVF): A review of 3 decades of clinical innovation and technological advancement. *Therapeutics and Clinical Risk Management, 2*(4), 355–364.

Wentzell, E. A. (2013). *Maturing masculinities: Aging, chronic illness, and viagra in Mexico* (p. 2013). Durham: Duke University Press.

Wexler, K. (1995). Egg-swaping scandal still unfolding. *Washington Post,* 10.

Yovich, J. (1994). Transabdominal gamete intrafallopian transfer. In J. G. Grudzinskas, M. G. Chapman, T. Chard, & O. Djahanbakhch (Eds.), *The Fallopian Tube: Clinical and Surgical Aspects* (1st ed., pp. 213–227, Chap. 16). Springer-Verlag London Limited. Retrieved from http://www.researchgate.net/profile/John_Yovich/publication/235935827_Transabdominal_Gamete_Intrafallopian_Transfer/links/02bfe5147c572c550f000000.pdf.

Zárate, A., & Basurto-Acevedo, L. (2013). Notas históricas sobre la investigación científica en el IMSS. *Revista Médica del Instituto Mexicano del Seguro Social, 51*(6), 650–655.

Part II
Reproducing Assisted Reproduction

ARTs as Technological Innovation and Cultural Novelty

Previously I had said that when I began studying ARTs in Mexico, back in 2006, I did not see them as "new" reproductive technologies. I claimed that calling them "new" was inaccurate because artificial insemination had been successfully used for over 100 years, IVF had its first successful outcome in 1978, and in Mexico, there had been more than 22,000 cycles performed since 1999 (according to the numbers offered by the RedLARA). Now, I want to claim that they are new, not in terms of their assemblage as *technological* artefacts but in terms of their assemblage as *cultural* artefacts, and I want to trace how these cultural novelties began to loose their "newness". Loosing "newness" means that the artefact becomes more familiar because it has been socialised, and maybe even routinised (Wahlberg, 2018). (The label "new" can also be defended if we consider that these technologies are in constant change; hence, in a way, they are perpetually new.)

Technological innovations usually are also cultural novelties, as such they require several different sorts of practices, for example research, design, production, introduction, diffusion, adoption, consumption, socialisation, normalisation, and routinisation (although these phases do not necessarily happen in a linear consecutive pre-established order)[1]. These practices imply creating and transforming political, economic, and social structures in order to make the technological innovation accessible and acceptable. Likewise, they imply creating terms, myths, and rituals that will help people incorporate them into a broader cultural framework. In this part of the book, I explore how assisted reproduction was made usable focusing on the practice of diffusion, socialisation, and normalisation that took place in the first decade of the twenty-first century.

Everett Rogers notes that "getting a new idea adopted, even when it has obvious advantages, is often very difficult [...] Many innovations require a lengthy period, often of some years, from the time when they become available to the time when they are widely adopted" (1983: 1). Louise Brown was born in 1978; forty years later (when I am writing this book), there are at least 8 million people that have been conceived through IVF. How much would that number (8 million) increase if we compute users who unfortunately did not achieve the desired child? Quantifying AR practices is a difficult thing; it requires that countries keep track of the number of clinics offering services, procedures performed, and outcomes. Not many countries have such a specific and comprehensive registry; Mexico definitely lacks it. So, numbers, statistics, and percentages given regarding assisted reproduction should be considered very rough approximations. In Mexico, in one year (2014), at least 9221 cycles were initiated; this number only represents the activity of 31 of the approximately 90–100 active clinics in Mexico (Zegers-Hochschild, et al., 2017). Nevertheless, these numbers can serve to think about whether AR has had a quick diffusion and hence adoption process among "patients" and practitioners or if it has been a slow one? How was the diffusion of ARTs among the different group of users?

---

[1]I speak of practices and not moments or stages because I see them as more than moments in time and place and because I want to highlight that these practices can be taking place simultaneously in different (or the same) settings and involving different (or the same) actors, they are ways of doing, of creating the technological innovation. They are the assemblage of entities (people, institutions, objects), interacting (doing something), producing outcomes (building world).

Who were these early adopters? Following the tradition of STS, I consider that users "come in many different shapes and sizes" (Oudshoorn & Pinch, 2003), they are more than just the patients, they are also the healthcare professionals, the journalists and screen writers, and the policy makers. Each of these uses ARTs in different ways: as a way of reproducing, as their professional field, as the topic for their fictional and non-fictional stories, as an object of regulation. When they "use assisted reproduction", they are contributing to their construction and usability, to their acceptability, and demand. In their different discursive productions (papers, presentations, articles, advertisements, stories, and laws), they are contributing to ART's mythology, its rituals, and its terminology.

Diffusion, as used by Rogers, points at "the process by which an innovation is communicated through certain channels over time among the members of a social system" (1983: 5)[2]. This communication can be done through various channels (mass media, interpersonal, public spaces, scientific publications, etc.), and its main objective is to create "awareness-knowledge" (Rogers, 1983: 18). After diffusion, after people have become aware of the existence of a new technology, comes the process of deciding whether or not to adopt it (what Rogers calls the innovation-decision process). But in order to be able to adopt a technology, it must first be accessible, acceptable, and there must be a need for it.

Many technological innovations are also cultural novelties (Oudshoorn, 1999), that is, they may lack a demand for them, they might not be fully socioculturally acceptable, and they probably do not have a set of terms, rituals, and myths that help people makes sense of them. Therefore, every technological innovation needs a cultural intervention to create a demand for it, generate its accessibility, and make

---

[2]Rogers clarifies that while some make a distinction between diffusion and dissemination, where the former is spontaneous and unplanned, and the latter is directed and managed, he uses both interchangeably because drawing a clear distinction in practice of how things happened is not possible. Since I see these as practices that are situated, complex, and involving a great variety of actors, I do not find a clear line between the planned and the unplanned.

it somewhat acceptable.[3] One element in the process of making a technological innovation usable is evaluating the innovation. Rogers identifies five points which relate to how the innovation is being perceived: (i) is the innovation perceived as offering an advantage, (ii) is it seen as being compatible with established values, (iii) is it thought of as too complex, (iv) does it allow for trails, and (v) how observable are the desired outcomes of the innovation (1983: 17). The practice of making usable is also a practice of creating the symbols, myths, and rituals that will help this technological innovation become acceptable, then normalised and routinised. It is in these practices that moral, aesthetic, political values are assigned to them. In this chapter, I show how these five points are put into action in order to make assisted reproduction usable.

In Chapter 5, *The Universe Is Expanding*, I see how assisted reproduction was made usable for practitioners and thus available for patients. I begin by looking at how they describe infertility and assisted reproduction in the books they produce and use as reference. Then, I describe how new clinics emerged and how the profession congealed by establishing links with the Latin American community through the participation in RedLARA. Many of the physicians participate also in the US association, ASRM, and in the European one, ESHRE. However, as opposed to the relationship with RedLARA, these relationships are seldom present at local meetings and conferences, and they tend to become visible in the clinic's webpages where they display their logos.

In Chapter 6, *The Discursive Landscape*, I show how assisted reproduction is presented as offering an advantage over adoption for those who are involuntarily childless because it allows future parents to maintain the biological link; where the complexity of the procedures is simplified by comparing them to the natural process; where they offer terms, create rituals, and build stories to help make assisted

---

[3]Accessible is not the same as affordable. As used here, accessible means that there is, at least conceptually, the possibility of being used. Accessibility is achieved, for example, when people know that these services exist in their country, when they are eligible to use them, and when they can pay for them either because they have the money or because they can use credit. Credit cards, for example, make things that were unaccessible, due to lack of funds, at least apparently accessible through a credit that allows one to spend money that one does not have.

reproduction compatible with Mexican culture; and where the desired outcomes (pregnancy and live healthy baby) are constantly presented. These spaces are part of a discursive landscape conformed by public areas (e.g. advertisements in public transport and in billboards), the media (press, television, radio), the Internet (websites, blogs, and social media), support groups, recruiting events organised by clinics, and a trade show. The narrative built by all these spaces offered users and potential users ways to make sense of the various ARTs, of how to manage the ethical, moral, philosophical, and cultural tensions they create, and of how to name things, processes, and relationships (in sum, how to make AR usable). These narratives were created between 1990s and 2010, which is when the early adopters were encountering and adopting ARTs.

References

Oudshoorn, N. (1999). On masculinities, technologies, and pain the testing of male contraceptives in the clinic and the media. *Science, Technology, & Human Values, 24*(2), 265–289.

Oudshoorn, N., & Pinch, T. (Eds.). (2003). *How users matter: The co-construction of users and technologies*. Cambridge: MIT Press.

Rogers, E. M. (1983). *Diffusion of innovations* (3rd ed.). New York and London: Free Press and Collier Macmillan.

Wahlberg, A. (2018). *Good quality: The routinization of sperm banking in China*. Oakland: University of California Press.

Zegers-Hochschild, F., Schwarze, J. E., Crosby, J., Musri, C., & Urbina, M. T. (2017). Assisted reproduction techniques in Latin America: The Latin American Registry, 2014. *Reproductive BioMedicine Online, 35*(3), 287–295. https://doi.org/10.1016/j.rbmo.2017.05.021.

5

The Universe Is Expanding

1994 was an election year; a year that turned out to be full of important sociopolitical events: Mexico was to vote for the last president of
the century, it was the last year of Carlos Salinas' administration, international agreements were signed, and local uprisings were settled. The
relationship with Vatican had been re-established, and certain laws
had been changed to allow for the public display of religious practices (de la Torre, 2008). After signing the North America Free Trade
Agreement (NAFTA) in 1993, the Ejercito Zapatista de Liberación
Nacional (EZLN) raised in arms on 1 January 1994 in San Cristobal
de las Casas, Chiapas.[1] Mexico, and the world, received the new year
with a juxtaposition of images: on the one hand images and videos of
men and women wearing black balaclavas declaring war against the government, on the other, images and videos of president Salinas sitting at
his presidential desk stating that this movement was against Mexico and

---

[1]One of the leading complaints of the EZLN was the signing of the NAFTA a few months
before, in 1993. Signing this agreement meant the abolition of Article 27 of the Constitution
which protected the native communal landholdings from being sold and privatised. Without this
protection, land could now be bought by whomever without limits, for example large food cor-
porations (Kelly, 1994).

© The Author(s) 2020
S. P. González-Santos, *A Portrait of Assisted Reproduction in Mexico*,
https://doi.org/10.1007/978-3-030-23041-8_5

that it was made up of national and international professional violent groups as opposed to indigenous groups, as they were claiming. A few months later, on 23 March, Luis Donaldo Colosio, president candidate running for the PRI, was assassinated. On 18 May, Mexico joined the Organisation for Economic Co-operation and Development (OECD). In December, Ernesto Zedillo became president, he would be the last president of the 70-year dynasty, because together with the turn of the century came a change in power.

During Carlos Salinas' presidency (1988–1994), the globalisation process accelerated, a "tsunami" of foreign influence was evident: consumer products, television programs, ideas of individuality, and consumerism. What had once been a political economy of import substitution and industrialisation projects was transforming into a political economy of imports–exports. A country with a population that Ricardo Hausmann qualified as "weak consumers" (Morton, 2003) had to be taught to spend, buy, dispose, and consume. A country that was accustomed to produce for its internal market had to change its ways to compete and satisfy the foreign consumer. A country that had spent years looking inwards for stability and growth was now, for the first time, associating a better future with better relations with Washington (Fuentes, 1993). Mexicans were asked to openly re-signify the USA, from the imperialist bully to the friendly partner (Preston & Dillon, 2004)[2]; they were asked to embrace and follow the American Dream.

All this influence was intertwined in the new set of family planning campaigns put out in the 1990s. The main slogans were now: *"Planifica, es cuestión de querer"* (Plan! it is a matter of wanting), *"Tú decides si te embarazas"* (It is your decision if you get pregnant), *"Hombres y mujeres, diferentes pero iguales"* (Men and women, different but equal), *"Porque tus desiciones son importantes, infórmate"* (Because your decisions are

---

[2]I say openly because, as pointed out by Preston and Dillon (2004) Mexicans have had a close relationship with the USA long before the NAFTAA large percentage of the population has had a direct relationship with the USA, because either them or their family members have worked or studied in the USA, because they shop for goods and services, or because they spend holidays in the USA.

important, get information), "*Infórmate, es tu derecho: Planifícatel, una buena opción*" (Get informed, it is your right: Planificatel,[3] this is a good option). In these slogans, family planning was framed as an accessible option; they emphasised values of individualism, responsibility, empowerment, human rights, and gender equality by stressing the fact that family planning was an individual's decision, a matter of will, and a good option (Nazar-Beutelspacher, Zapata-Martelo, & Vázquez-García, 2004: 148). At this stage, the messages targeted a broad audience including teenagers, young adults, families, and elder people, from both rural and urban settings. Not only did they seek to promote family planning values, but also awareness about sexual and reproductive health and about the services offered by the government (e.g. planificatel). It was also in this period that the first male-oriented family planning campaign was launched: Vasectomía sin bisturí (vasectomy without scalpel), although it had little impact (Gutmann, 2009).

Between 1988 and 2000, Mexico's total fertility rate went down from 3.7 children per woman to 2.4 (CONAPO). This decrease was accompanied by a shift in the government's perspective from population control through family planning to reproductive health. Changing the emphasis from the family planning paradigm to the reproductive health paradigm meant shifting attention towards the "institutional, cultural and political context in which the decisions pertaining to reproductive and sexual behaviour" take place (CONAPO, 1999: 185). This new paradigm brought together ideas and policies on reproductive rights, family planning, maternal and child health, infertility, and STI-D. A new programme called Programa de Salud Reproductiva (Reproductive Health Programme) was created; its purpose was offering information and orientation regarding prevention, diagnosis, and handling of infertility. Family planning campaigns and the new reproductive health perspective were spreading the message that, through biomedical knowledge and technology, it was possible to manage and control reproduction, i.e. assist reproduction.

---

[3] *Planifícatel* was a telephone helpline for family planning issues. The name is the merging of 'Planifica' which means to plan and 'tell' as in telephone.

However, not only did the fertility rate decrease, but also the population pyramid began to flip. The effects of the family planning campaigns as well as of the hygiene and healthcare campaigns were starting to be felt. The elderly population was becoming larger and the young population smaller, affecting how resources were generated and distributed: there was an increase in the number of pensioners and a decrease in the number of contributors. Likewise, due to changes in the epidemiological profile of the population, chronic-degenerative ailments became more common. However, the infectious diseases did not disappear entirely; in fact, they remained an important health concern in particular areas of the country. This burden was made heavier with the neoliberalisation of the healthcare market. In brief, the type of neoliberalism installed in Mexico since the administration of Miguel de la Madrid sought to defend the individual's liberty, protect the market, favour private investments and industry, and discourage the welfare state, the protection of public goods, and the improvement of public services. During his presidency, investment in health was reduced in a 47% then, with Salinas and as part of the NAFTA, federal laws and regulations were amended to allow for international private insurance companies to enter the national market. These reforms to the healthcare system, approved in 1995 and put into action in 1997, helped the private medical sector to grow and expand (Tamez González, Eibenschutz, Camacho, & Hernandez, n.d.; Tamez González & Valle Arcos, 2005).

During this time, the field of infertility and assisted reproduction went from being a research topic and a highly specialised procedure offered at a few healthcare services, to becoming a fully established industry. After reporting the first successful outcomes between 1988 and 1991, and once the first generations of professionals were graduating from the speciality programmes offered both in Mexico and Spain, a new cohort of AR services and clinics began to appear across the country. The number of clinics increased from two in 1986 to approximately 25 clinics in 2003. Improvements in the tools used both in the labs (to care for gametes and embryos) and the clinical setting (to care for patients) created the opportunity for new procedures, such as ICSI, sex selection, and PGD, to become standard practice displacing older ones, as was the case with ZIFT and GIFT. Worldwide there were now "thousands of children born as a result of assisted reproduction" (Pérez Peña, 1999: 249). The community of national professionals were coming into contact and collaborating with practitioners from across Latin America. The objectives and perspectives of the association were undergoing important transformations, to the point that, in 1993, during the administration of Carlos Hinojosa, the association changed its name again, now to AMMR (Asociación Mexicana de Medicina de la Reproducción, Mexican Association of Reproductive Medicine). This new emerging industry involved now support groups for users (e.g. Asociación Mexicana de Infertilidad), marketing strategies, financing options, a magazine (Tu Fertilidad), a yearly expo (Expo Fertilidad), some media coverage (within national television and radio programmes), ethical and legal debates within the academic field, and some efforts to regulate practice. All this suggests that by the year 2000, the Mexican AR system had reached a new phase in its development. The field both nationally and internationally was much more solid than in the 1950s or even the 1980s.

In Part one, I followed the dawn of a new era in the caring for infertility, an era focused on the reproductive technologies that manipulate gametes outside the body. In this chapter, I will look at how this new way of caring for infertility was diffused and eventually accepted by physicians and embryologists. I focus on how the universe of assisted reproduction expands. The chapter is guided by the comments of two former presidents of the association. The first appeared in the introduction of a book edited in 1999 and the second on the reedition of this same book, published in 2003. One saw the gradual changes that have affected the objectives of the association while the other saw the revolutionary changes in the area of reproductive medicine:

> In the fifty years of our association, we can observe gradual changes in terms of the objectives for which it was created. (Kunhardt Rasch, 1999)

> In the last years there have been revolutionary advances in the area of reproductive medicine… (Téllez Velasco, 2003)

Diffusing Knowledge Among Practitioners

Articles, books, and conferences are tools through which new followers sought to be enrolled into the field (Callon, Law, & Rip, 1986); hence, they are useful sites to study how a knowledge set gets disseminated. In this section, I look at conferences, articles, and books that were produced between the mid-1980s and the 1990s and played an important role in diffusing the knowledge of assisted reproduction among medical students and medical practitioners.

As mentioned before, the first groups to work on gamete manipulation were located in public institutions, first at the National Medical Centre and then at the INPer. Because these were research and teaching institutions, their members were asked to publish their findings (usually in Spanish and in journals of easy access), to organise conferences (to which national and international speakers were invited), and to publish the papers or at least the abstracts. Furthermore, the clinical work done at these places gave students the opportunity of learning hands-on about these procedures. All this contributed to disseminating the knowledge about ARTs:

> We began to publish, and people in Guadalajara, Monterrey, San Luis Potosí, even in Mérida, they were reading what we were publishing, this inspired them. Obviously people went to many places for training–the USA, Spain–but I do believe that the inauguration of the clinic [INPer], and the expansion of the knowledge that came out of this clinic, that came out as publications, conference presentations, etc. [...] because if AGN and IECH would have been the only ones to exist, nobody would have known that, in Mexico, there were people doing this. (Kably, Interview, 2017)

INPer's annual conference frequently included presentations dedicated to assisted reproduction, covering topics such as surgical procedures and IVF (in 1987), sperm preparation, the ethology of infertility, and IVF-GIFT (in 1988). The attendance of these annual conferences was, in average, 500 people. National and foreign experts were invited (e.g. Ricardo Asch), and the abstracts of these presentations were then published in the

Journal Ginecología y Obstetricia de México. During the fifth annual meeting, a full course on the "Advances in Human Reproduction" was organised by Kably and Delgado Urdapilleta. The memories of the course were subsequently published. The course included talks on artificial insemination and sperm preparation, anatomy of the fallopian tubes and tubal gestation, infertility related to endometriosis, ovulation issues, and chlamydia, surgical procedures, laparoscopy, and hysteroscopy, frequent miscarriages, and IVF. It is worth highlighting that, although GIFT was a technique that they were using in clinical practice, that was developed by Ricardo Asch—whom they frequently invited to give talks and courses—and that Mexico's first two successful cases were achieved using GIFT, it was omitted from the publication (according to the list of abstracts published in 1 Ginecología y Obstetricia de México, it was covered during the oral presentation). True, the event took place in April, just a few weeks after the first two birth were reported (the first had been in February and the second in March); however, conferences are precisely the place where scientists present their latest results, and this particular course was entitled "Advances in Human Reproduction", so it is noticeable that there was no mention of this technique nor about these firsts. Instead, there was an emphasis on IVF, particularly in the expansion of its usability, presenting it as a procedure that could be used beyond cases of bilateral tubal obstruction, "the indications for IVF are expanding as the technique has become more successful" (Kably Ambe, 1988: 114). In interviews, I have asked physicians to tell me why they decided to move away from GIFT and towards IVF; their common response has been that IVF offers them the possibility of ensuring fertilisation while GIFT does not.

Publications require the coordinated and collective work of interested people that see an advantage in undertaking such endeavour. In 1959, Arzac had the idea of editing a collective book, *La Esterilidad y la Infertilidad Humana*, with the purpose of giving an account of the knowledge and experience accumulated by the association during its first ten years. He invited 25 authors to collaborate with him in writing and coordinating a total of 45 chapters. Many of them never handed in their manuscripts, so the project had to be abandoned. Then, in 1999, Efraín Vázquez Benitez (president of the AMEE from 1977 to 1978) tried this again. The occasion was the association's 50th anniversary.

He was able to secure funds from the pharmaceutical company Serono to help finance the publication, and to convince several former presidents of the association to help him coordinate the book. Each was in charge of one section and of commissioning the chapters to national experts in the field. The result of these efforts was *Medicina Reproductiva*. Due to the "unexpected success of the first edition" (Téllez Velasco, 2003, Presentation) among professionals and students, it was reedited in 2003. Analysing the two editions of the book allows us to identify how this community of professionals were building their history, and to track the way the field was changing epistemologically, technologically, and morally.

Community

In these books, we can see how this group of professionals were building their community and establishing their identity by claiming their past, present, and future. In both editions, they set their link to the past in several ways. First, they link to the distant past through the Editorial Comment written by Vázquez Benítez and the Introduction written in 1999 by Kunhardt, where they frame this book as a commemorative publication due to the 50th anniversary. This link is evolutionary (i.e. through change); they indicate that although there have been changes, these have not been fundamental enough to demarcate their relationship with the association founded in 1949. Furthermore, they underline the importance of illustrating or leaving evidence of what reproductive medicine has been and now is. Second, they also link to the distant past by dedicating the second edition to the eleven former deceased presidents while also linking to their recent past by inviting the living former AMMR presidents to coordinate each section. Third, they link to the past and create a sense of community by repeating their origin story. In the Editorial Comment, Vázquez Benítez repeats the story of those "34 restless professionals who initiated an organised scientific exchange focused on investigating, understanding, controlling, and promoting many different aspects of human reproduction [...] hundreds more of active members have joined since" (Vázquez Benítez,

1999, Editorial Comment). Constantly retelling this story in journal articles, book chapters, and conference talks, regardless of the variations of the exact date and number of people involved, gives it a foundational power, it becomes mythical.[4]

Remembering the forefathers is one way of linking the past with the present ('he was my mentor') and the present with the present ('we share the same mentor'); another way of establishing and strengthening ties is by granting people specific roles and special places within the community. Not only were former presidents of the association asked to coordinate the sections, but they were individually selected "based on their experience and views and they decided the content of each chapter" and they were given the responsibility of selecting who would write them. Emphasising the point about their "experience and views" distinguishes each individual based on their special knowledge and experience. This gave these people the power to guide the course of the field, to decide what is said, and who said it.[5] In both editions, the section on assisted reproduction was coordinated by Efraín Pérez Peña[6] (president from 1995 to 1996). Certain authors covered the same topics in both editions; for example, Gutierrez Najar wrote about the Blastocist, the IECH team wrote about IVF, Félix Arce and García Ramirez (both from the Creasis San Pedro clinic est. 1996) wrote about GIFT, and Gutierrez Gutierrez, Perez Peña, and Gallardo Lozano (all from the Instituto Vida clinic) wrote about ICSI. Two things can be highlighted from this. First, in the authorship we see how teams are made; with the exception of Gutierrez Najar—who did not write with his team—the rest of the groups of authors were co-leaders at their clinics. Second, people were not writing on the issue they had been credited with being "the first" to achieve success—IECH with achieving the first birth after a successful GIFT, Gutierrez Najar

---

[4]This story has also been repeated in, for example, López de Nava 1999 (Book: Medicina Reproductiva); Vázquez Benítez (2008) (Journal: Reproducción).

[5]These articles were written by the leading national AR specialists, most of which worked at AR clinics located in Mexico City, Monterrey (NL), Guadalajara (Jalisco), León (Guanajuato), Matamoros (Tamaulipas), San Luis Potosí (San Luis Potosí), and Puebla (Puebla).

[6]He is founder of the chain of clinics Instituto Vida, he works in Guadalajara, Jalisco.

with achieving the first IVF baby and with being a founding member of RedLara. Moreover, these "firsts" are not even clearly mentioned. In 2003, new authors were incorporated, one of which was the embryologist responsible for the first successful GIFT procedure in Mexico, Genaro García Villafana, however he is coauthoring the chapter on IVF and not on GIFT. If by repeating the story of the formation of the AMEE it acquires force, what happens to the stories of these firsts births when they are only vaguely mentioned or completely omitted in texts like this one?

As mentioned above, these books also worked as tools to build a specific sort of identity for this community. These books show the association's strength, endurance, and establish their relevance by displaying some of their scientific achievements. They were conceived as a "compilation of the thoughts and achievements" of the group of AR professionals and emphasised that their accumulated experience was "not the result of pure chance, but the product of many years of work" (Vázquez Benítez, 1999, Editorial Comment). They highlight their aim of "moving beyond a mere biotechnological description of the field, regardless how spectacular it is, and come closer to an integral humanistic perspective" (Vázquez Benitez, 2003: XXVI). This move suggests no longer focusing only on the spectacular nature of the biotechnology, but on what this biotechnology can do and how to think about it, highlighting the importance of engaging with the ethical problems these technologies generate.

In addition to linking with the past and expressing interest in the present, a third way in which they establish their community is by looking into the future; proving that its field and community are solid enough that they will live on into the future. This is done in these books in two ways: first, by including a reflection about what will come in the near future and how to prepare for it; and second, by the increase in participation. For a community to continue its existence there needs to be interest to do so. Analysing and comparing these two editions, we see an increment in participation. The first edition had 76 authors, eight of which were women, writing a total of 78 chapters, gathered into 20 sections. In 2003, these numbers went up, there were 110 authors, of which only 17 were women, writing a total of 91 chapters gathered into 19 sections. The increase in chapters and in authors could reflect the increase in the size of the filed in terms of number of professionals working in the field and in the number of procedures that make up the field.

Epistemic

These books had the objectives of offering a panorama of reproductive health in Mexico and an overview of the "state of the art" of the field. Simply by looking at the different sections and chapters, we can get an idea of the thematic conformation of the field, a sense of what the filed "is made of", of its epistemic and moral composition, and of how this has changed. Although the books covered a wide range of topics related to reproduction,[7] not all topics received the same degree of attention. As indicated by its editor, Vázquez Benitez in 2003, this unbalance "is a parameter of the current interest" in each of these areas.

In both editions, the section on assisted reproduction was the largest within the book. While the average was five chapters per section, the section on assisted reproduction had 15 chapters in 1999 and 20 chapters in 2003. One possible reason for this unbalance could be that by, the turn of the century, assisted reproduction had become *the* area of interest. But what exactly was assisted reproduction at that moment? What did it mean for the members of the association?

---

[7]Between the first and the second edition, there are three important changes in terms of the book's structure. The first change is regarding the sections, three new sections were added to the 2003 edition: "Recurrent Gestational Loss", "Conventional Human Reproductive Surgery and Endoscopy", and "Other Situations that can Affect Reproduction", this last one with sections on cancer, life style, ectopic pregnancies, and genetic factors. The rest of the book's chapters were the same. Both included sections on the different reproductive periods (from puberty to menopause), on family planning (contraception and sterilization), on infertility (note that there are eight sections dedicated to female factors and only one to male factors), and on assisted reproduction. Both also included chapters on education and reproduction, and on the legal and bioethical aspects of intervening in reproductive processes.

Assisted reproduction denoted "all those procedures aimed at achieving pregnancy that include manipulating gametes outside the body" (Pérez Peña, 1999: 249). It encompasses a range of techniques and procedures that included ovarian stimulation, artificial insemination, IVF, ICSI, pre-implantation genetic diagnosis, fertility preservation, and a variety of laboratory procedures and tools.[8] These are procedures that have been underdeveloped for the past 100 years and that still today are undergoing constant transformations and re-inventions. They are procedures that are situated within a wider biotechnological scenario. For example, the announcement, in 1996, of the successful cloning of Dolly the sheep or the way the Human Genome Project (1990–2003) was carried out. Particularly, the way the two research groups raced towards completion, the international and public group and the Celera Genomics group, and the different political positions they held regarding knowledge production and ownership. These projects sparked global ethical debates and concerns regarding the political and commercial aspects of gen-related technologies. One example of these debates was concerning the techno-scientific and commercial paradigm followed out by Celera Genomics, a paradigm based on a particular way of sequencing genomes and on the premise that genes can be patented (e.g. breast cancer genes). The sum of the idea that genes are responsible for X condition (genetic determination), the claim that these genes can be found and manipulated, and the use of IVF resulted in the idea of manipulating these genes to eliminate undesired traits. Genetic editing, in the beginning, caused great concern with certain groups of people. For example, after the publication of a series of articles claiming to have found genetic alterations related to deafness, the community of deaf people became very concerned regarding what this will mean for them. Particularly if we take into account that they do not consider deafness to be a health issue. This drew attention to the

---

[8]For example, assisted hatching, the use of fluorescence in situ hybridization (FISH), polymerase chain reaction (PCR), cryopreservation of embryo, improved visualisation instruments, faster and more accesible hormone determination techniques, criteria to evaluate sperm and embryos.

question of how determinant are our genes? In the 1999 edition of the book, Álvarado Duran followed a deterministic perspective: "genes determine all aspects of health, illness, and behaviour" (p. 389). The predominance assigned to genes has helped justify certain procedures in the field of assisted reproduction. For example, experimental techniques such as nuclear and cytoplasmic transfer, "used in cases where the woman is of advanced age and insists on having a genetically related child [...] as well as for cases with mitochondrial problems" (Pérez Peña, 1999: 278). Another important field being explored at that time was gene therapy. USA-based researchers were working on this. John Zhang[9] and James Grifo were working on spindle nuclear transfer, and Cohen was performing ooplasmic transfers. Cohen's work had resulted in the birth of over a dozen children; however, safety concerns lead to a ban of these techniques in 2001 (Cohen & Alikani, 2013). These concerns were shared by the authors of this chapter, they too pointed out that while these techniques represent "a great advancement, their security and efficacy are not yet well determined, more research needs to be done to assess if the risks outweigh the benefits" (p. 279).

These two editions show how the developments in ARTs have influenced shifts in the objectives of the field. In 1999, Váquez Benítez still described the interest of the association as being "sterility, infertility, and contraception"; by the 2000s, the areas covered by reproductive biology became menopause, contraception, maternal-foetal health, and assisted reproduction. The change in wording (from sterility-infertility to assisted reproduction) is important. First, the focus was on the condition and its cure, then the emphasis was on the procedures, ARTs as the techno-fixes for situations that go beyond biological. In the 2003 edition, there was a chapter on the Indications for Assisted Reproduction which ended with the following paragraph that clearly illustrates this shift:

---

[9]Zhang had been banded from China, his native country, due to his work on spindle nuclear transfer. In recent years, 2016, he achieved the first birth using mitochondrial replacement techniques, a procedure he allegedly conducted in Mexico (see González Santos, Stephens, & Dimond, 2018).

It seems like there are more indications for these techniques. In fact, we could conclude that, currently, the selection of the technique responds more to the results and complications of the technique, than to the diagnosis. This is how the use of procedures such as GIFT, ZIFT, PROST, TET, etc. is decreasing and instead the use of procedures like IVF and ICSI is increasing. The ultimate goal is still to improve pregnancy rates, and reduce the complications and the multiple pregnancies, and hoping to have a new born at home. (Serviere Zaragoza, 2003: 289)

However, this shift was not performed without tension. Just a few pages further ahead, in the chapter on Counter-Indications for Assisted Reproduction, which was one of the chapters added in 2003,[10] the author states:

The first counter indication that should be considered is the lack of indication. It is now ever more frequent to find clinics suggesting IVF or ICSI to couples who do not fit the criteria, or even in cases in which they have not even been evaluated. [...] This stems from the great laxity with which people are managing assisted reproduction techniques and the lack of regulation these technologies have, and the lack of adherence to a generalised consensus". (Tlapanco Barba, 2003: 291)

Tlapanco Barba points out that professionals are displaying a "great laxity" with how they manage ARTs, a "lack of adherence" to the consensus, and how this is happening within the context of "lack of regulation". The laxity he mentions is because some still saw assisted reproduction as having limits: assisted reproduction is not "the panacea nor the first option for infertility cases, they have very specific indications" (p. 275). However, these limits were becoming ever more flexible, the criteria for inclusion was now allowing for the inclusion of more cases as eligible for these techniques. For example, egg donation was originally indicated in cases of women of "advanced age, ovarian failure, or a genetic condition", its use was then expanded to include cases of women with low response to the ovarian stimulation drugs, or repeated

---

[10]The new chapters in the 2003 edition were on indications and counter-indications for AR, the myths on sex selection, PGD, and on the use of hormones (e.g. FSH, LH, and hCG).

IVF failure. This last indication—repeated IVF failure—is no longer a diagnosis pertaining to a problem (pathology) related to the anatomy or physiology of the reproductive system of either the woman or the man, it is simply a failure in the procedure. This suggests, as said in the previous quote, by Serviere Zaragoza, that the "selection of the technique responds more to the results and complications of the technique, than to the diagnosis". Because the definition of infertility has remained unaltered since the 1950s. They still frame infertility as a medical condition diagnosable only in heterosexual couples: not achieving conception after twelve months of regular unprotected sexual intercourse.

This expansion had ethical and moral limits. For example, there were concerns related to health matters, particularly with the possible consequences of using, for example, poor quality sperm during ICSI. Their suggestion was to perform genetic studies on men whose sample requires the use of ICSI. Another limit was regarding the use of ARTs in certain cases in which reproduction was seen as questionable; for example, same-sex couples and single women (single men were not even though of as potential users). In the chapter on the indications for IVF, the authors (Galache Vega et al.) state that a marriage certificate has to be presented as part of the evaluation for eligibility (2003: 365, 376). In 2019 there are still many practitioners and policy makers that consider that the use of ARTs should be limited to married heterosexual couples, however there is a growing number of clinics that are now catering for same-sex couples and single individuals. This change of perspective could be the result of many factors, from the increase in demand and a legal backdrop (laws that now make same-sex marriage possible) that makes offering these services to same-sex couples legally easier, to a growing awareness and acceptance that people in all sorts of relationships can want a child and can offer a family context for this future child.

Finally, another important change that I have been tracing is regarding the place assigned to adoption with the realm of ways of solving the problem of undesired childlessness. Both editions have a chapter dedicated to adoption, however in 1999 it was a stand-alone chapter and in 2003 it became a chapter within the section on "Sterility of Unresolved Cause", thus framed as an option only when all other options have not revolved the state of childlessness.

Normative: Morals, Ethics, and Laws

In both editions, Manuel Álvarez Navarro coordinated the section on Bioethics and invited Efraín Vázquez Benitez to write the chapter on Science and Bioethics, Alberto Alvarado Durán to write the chapter on Bioethics and assisted reproduction, and Carlos Fernandez del Castillo to write on Bioethics and Contraception. Likewise, the section on the Legal Aspects of Assisted Reproduction was twice coordinated by Juan Rodriguez Argüelles. Thinking about the moral and legal implications of intervening in reproduction was something present in this association since its early stage, we can find articles addressing this issue in Sterility Studies. However, the situations in which they now had to make decisions were very different to the ones present in the 1950s. The scenarios and possibilities facilitated by the different high complexity ARTs (i.e. IVF) lead to new sorts of questions; after all, artefacts have politics. In a very broad sense there were four issues they were struggling with (enlisted here not in order of importance): whether to use or avoid the techno-fix, the biopolitical role of the physician, how to re-make sense of cultural notions and ways of the past in light of the new scenarios (e.g. kinship, marriage, and sex), and how to make ARTs useable to them, as physicians.

Álvarez Navarro and Vázquez Benítez worried about the effects the changes in the world were having on the moral and ethical make-up of the physician: "We cannot deny the effects that some recent events have on the integrity of the physician" How were physicians going to face the changes that were happening and that were coming ahead? How can they solve their ethical and moral conundrums in order to be able to accept (or not) to practice this sort of medicine? These authors recognised the limitations of the four pillars of bioethics which, in general, guide medical conduct: respect for autonomy, non-maleficence, beneficence, and justice. Instead, Vázquez Benítez offers four points to think about which I use as guides to look at how they were thinking about the moral, ethical, and legal aspects of assisted reproduction.

First, they advised to be careful with the tendency to think there is a magical formula to solve all issues:

medicine has been a singular field where technique and technified instrumentation have flourished [...] while medicine has benefited from the opportunities offered by the innovations in technological science, it has also fallen to excess in resorting to technology for every solution, to every health problem. Science and the medical art seem to have divorced due to an irreflexive veneration to the technological presence. (Álvarez Navarro, 1999: 379)

As indicated in the quote, there was a concern with the uncritical way physicians were adopting the technological fix and with which they were trying to solve moral and ethical issues. They stress the importance of establishing a distinction between technological feasibility (what can be done) and social acceptability (what we want to be done). In this sense, they ask if everything that technically can be done should be done or is there an uncritical fascination with the technological fixes.

Second, they highlight the difficulty of dealing with many of the entirely new problems and situations that are emerging as the result of the use of ARTs:

Establishing models of moral conduct within medical practice moves in two extremes: one faces the need to reaffirm the values and norms of conduct with a universal and immutable nature, on the other hand, there is a need to attend the ones that emerge from the inevitable human transformation [...] which demands adaptations and new prepositions [...] to which he ask: is the contemporary human being different from the one who lived centuries ago, or is there an immutable human dignity that outlives time? (Álvarez Navarro, 1999: 379)

Reaffirm the old values while attending the emerging situations for which these old values many times no longer apply, that is the problem these physicians (and society in general) are facing at the turn of the century. For example, questions regarding the purpose of marriage, the relationship between sex and reproduction, the way kinship can be established, who can reproduce, and whether if society should accept (or not) the new and different family confirmations. Vázquez Benítez added to these interrogations further ones regarding whether if all couples have the obligation and the desire to reproduce and if they all should reproduce (Vázquez Benitez, 1999: 384). Álvarez Navarro and Vázquez Benitez suggest that these problems become aggravated when trying to use out-of-date moral codes, when judging without fully understanding the situation, and when the relativity of morality is considered and thus there is no longer the belief that there is an immutable universal human dignity that outlives time.

> [...] seeking guidance in generic principles found in old codes to practical cases derived from the application of technologies that are constantly progressing [...] from passing moral judgement over situations for which we lack information and an impossibility of predicting the future (how many lives have been saved using procedures that once were objected to). (Vázquez Benitez, 1999: 384)

Third, the importance of recognising that infertility goes beyond being simply a biological problem; it is a multifactorial problem. Patients seek the physician's help in reproductive matters not because they are facing a life-threatening biological problem, as it is commonly understood, but because they are living a human problem, one that is inserted in a historical appreciation for reproduction and maternity (and where infertility has been historically also seen as a disgrace, even a punishment). Therefore physicians need to see their patients and their conditions "not as a biological problem to solve, but as a human problem, as a whole" (Vázque Benítez, 1999: 383). However, they highlight that in some cases, there might be a conflict of interests between the couple and society, for example with the allocation of public funds to cover for these expensive procedures (distributive justice). Considering infertility to go beyond biological matters can then be used as an argument to broaden the eligibility criteria so ARTs can be used for more purposes and in more situations.

The fourth suggestion he makes is to "[...] not only consider the interest of the infertile couple but also the interests of the future child" (Álvarado Durán, 1999: 388). However, by doing this a new order of conflict can emerge, the conflict of interest between the couple and the well-being of their future child, particularly when the well-being of the child is jeopardised due to the use of assisted reproduction. For example, cases in which there is the use of third-party gametes, when there is a medical condition that is obstructing fertility and that is considered inheritable, or simply due to the use of the procedures. This reflection made them think about their own role in this matter. If their professional aim is to achieve the birth of a healthy baby, what to do when this goal is at risk? Are they entitled to intervene?

Regarding the legal risks of practising AR, Alvardao Durán goes deeper into the three major malpractice situations commonly applied to medical practice—negligence, lack of skill, and deceit—to see how these apply specifically to assisted reproduction. He points out that, although it is not considered a legal problem, it is unethical to use the success rates of other clinics instead of your own, since this might raise the expectations of the patient. Nevertheless, this deceitful practice has grown and become quite common in the 2010s as the result of some clinics using more active and aggressive publicity campaigns. It is frequent to see websites, Facebook pages, and adverts using statistical data produced in other contexts, being unclear about where and how this information was produced, and how it should be read and used. He suggested there should be regulations concerning this practice; however, of the more than twenty proposals presented to Congress between 1999 and 2019, not one has included an article referring to this.

A growing concern among Mexican physicians was the practice of defensive medicine: "There is a definite tendency towards the increase in lawsuits against the institutionalised medical services and the individual professionals" (Rodriguez Argüelles, 1999: 403). One way of protecting themselves against lawsuits concerning the diagnostic and therapeutic practice was keeping track of everything they did and following the procedures accepted by the community of experts. For this, the practician should always be up to date by attending courses and conference "[…] to prevent legal medical responsibility […] the specialists not only should be up to date in his field but he should keep record of these activities, and he should follow the norms". These actualisations and records of actualisation are what the medical colleges do. In 2003, the COMEGO (Mexican College of Gynaecology and Obstetrics) won over the Endocrinology college the right to certify physicians in the area of reproductive biology. By this point, reproductive biology was considered to cover four areas: gynaecological endocrinology, family planning, menopause, and assisted reproduction. (Note here that they are now talking about assisted reproduction, no longer esterilología nor simply "procedures to treat infertility"; hence the name of the knowledge set has also clearly changed and the new one has established.) Both family planning and assisted reproduction include procedures that require surgery. However, neither internal medicine nor endocrinology covers surgery. This was the argument used to convince the CONACEM (Mexican Council of Medical Specialties) that reproductive biology should be certified by the COMEGO: "The people at the INNSZ, the endocrinologists, also wanted to certify reproductive biology, but we went to the CONACEM and argued that they are not qualified to perform surgery, which we are, so they cannot do or evaluate all the areas that the speciality requires. They were doing low complexity stuff, like programmed coitus or ovarian stimulation, but not laparoscopy" (Kably, Interview, 2017).

The last step suggested in order to reduce the possibilities of lawsuits was obtaining consent from the patient. Although the notion of "informed consent" as used today is not present in these texts, they do speak about of the importance of explaining to the patient their diagnosis, the risks entailed in the diagnostic procedure, the characteristics of the treatment, alternatives to the proposed treatment, the expected success rates, and to be sure to answer all questions: "informed patients follow the treatments better and have better outcomes [...] they make more informed decisions, have less complaints and demands afterwards. [...] In other words, their expectations are more attuned to their possibilities and they accept these better" (Rodriguez Argüelles, 1999: 405).

As mentioned in various occasions already, AR professionals have always considered it important to have their practice regulated by the law, not only by their own professional bodies. In 1999 Alvarado Durán reports of a combined effort to develop regulations for the AR services, bringing together representatives of the government and of medical associations. This regulation would be part of the section on the disposal of human tissues, cells, organs, products, or cadavers, of the General Health Law. They were going to incorporate entities such as the embryo and the pre-embryo. They also suggested the creation of an AR Committee, in addition to the Ethics committee, to overlook the way these procedures are carried out. They suggested limiting the applicability of these procedures to heterosexual couples that were either legally married or in a stable relationship (and neither of them is married to another person). They set a time limit of five year to the cryopreservation of embryos and if they were given up for adoption, the identity of the ones adopting and the ones giving up for adoption must remain indefinitely anonymous. They prohibited embryo reduction, the use of gametes for purposes other than reproduction, mixing sperm or ova of different origin, surrogacy, cloning, producing chimeras, and non-therapeutic genetic manipulation. However, this regulation never made it into the legal corpus, these articles and amendments were never incorporated. Between 1999, when the first regulating proposal was presented by Gonzalez-Martinez from the Green party (Partido Verde Ecologista de México, PVEM), and 2010, there were 20 proposals to regulate assisted reproduction, yet none ever passed (see annexe for a list).

The section ends by highlighting how no country has ever passed a law in this matter without first having the professional community agreed upon the ethical guidelines. Maybe this was why no proposal had ever been passed. However, in 2012 members of the community of AR professionals published the Mexican Nacional Consensus of assisted reproduction (Kably Ambe et al., 2012). This document was created through the collaboration of several members of public and private clinics. It did not include, however, the collaboration of embryologists or nurses, who are central in the daily operation of the procedures. The consensus covered seven areas: patient eligibility, ovarian stimulation, egg retrieval, embryo transfer, post transference supplementation, cryopreservation, and informed consent. However, having a consensus has not changed anything regarding the passing of regulating proposals. In 2019, when this book was finished, there still was no specific set of articles or laws concerning the use and practice of assisted reproduction.

Within the first years of Vicente Fox's administration, three governmental bodies related to assisted reproduction were created: CENATRA (in 2000), COFEPRIS (in 2001), and CNEGySR (in 2004). Centro Nacional de Transplantes (National Transplant Centre, CENATRA) deals with issues related to organ and tissue donation and transplants, and has taken over the task of looking over gamete and embryo donation. Comisión Federal para la Protección contra Riesgos Sanitarios (Federal Commission for the Protection of Sanitary Risk, COFEPRIS) is in charge of issuing the licenses and permits for healthcare establishments that deal with surgical and obstetric procedures and handle organs, tissues, and cells; they are the only official regulating body that, to some extent, looks after the establishment of laboratories in which gametes and embryos are kept and manipulated. The Centro Nacional de Equidad de Género y Salud Reproductiva (National Centre for Gender Equity and Reproductive Health, CNEGySR) was created with the purpose of dealing with gender equality and reproductive health issues.

Up to now, I have focused on how assisted reproduction has developed, disseminated, and become acceptable within the group of Mexican physicians dealing with issues related to infertility. Now, without loosing Mexico from my focal point, I broaden my scope to see what is happening in Latin America. In general, Mexico does not seem to have extremely tight relations with the rest of Latin America, but the field of AR is an exception. Although physicians turn to the ASMR and ESHRE as references, the only association that has had constant and active presence (i.e. at conferences, giving courses, and certifying clinics) in the Mexican field is the Latin American Network of AR, RedLara.

From Registry to Network: RedLARA

"El registro Latinoamericano es también de México, hay que entenderlo así". (Zegers-Hochschild, annual AMMR conference, 2017)

In 1990–1991 a group of Latin American AR specialists, among which was Alfonso Gutierrez Najar, led by the Chilean Fernando Zegers-Hochschild, organised the Latin American Registry of Assisted Reproduction: a consolidated database with voluntarily shared information regarding the procedures used by the subscribing clinics and their success rates. A compendium of this information was published annually. The first of these reports included information from 21 clinics from eight countries, the only Mexican clinic represented in that first registry was INPer[11] (Zegers-Hochschild & Prado Aravena, 1990). Data was collected and reported in a standardised way and sent to the headquarters via fax. The final compendium was distributed through the network's website. In 2010, RedLARA implemented an online system to upload the information case by case, and if wanted, as the cycle progressed. That same year, these reports began to be published in two journals simultaneously, the Reproductive Biomedicine Online (RBMO) and the Brazilian Journal of Reproductive Medicine (JBRA). Changing the publication platform modified the way the report was constructed and presented. When it was published by the network's website only, it was a lengthier publication, written in Spanish, with a user-friendly design, with a table of content, a list of all the many tables presented, and with more information. It included information about the procedures and methods, and it offered more correlations between elements. When they began publishing these reports in the two academic journals (RBMOL and JBRA) these reports had to comply with the word and table limitations as well as with the format guidelines set by the journal, and it was now written in English. The purpose of the registry and its reports was to create an instrument to allow clinics "to legitimate the efficiency of these treatments in a scientific way [...] create an educational tool that, together with health professionals, would allow couples to evaluate the cost and benefits of ART procedures; [...] develop a comprehensive regional database to serve as an external reference for each centre's self-evaluation; and [...] for epidemiological research" (Zegers-Hochschild, 2003: 356). Surely publishing in two international journals gave these reports a broader visibility, but was there something lost in exchange?

In 1995–1996, the registry expanded into a network: the Red Latino Americana de Reproducción Asistida (Latin American Network of Assisted Reproduction, RedLARA) and their goals also expanded.

---

[11]The countries that participated were: Argentina, Brazil, Colombia, Chile, Ecuador, Mexico, Panama, and Venezuela. In the report, they state that there were 21 centres, however, in the list of clinics there are only 19 names (Zegers-Hochschild & Prado Aravena, 1990).

Now they were interested in participating in the scientific, academic, and ethical development of the field within the region. To meet these goals, RedLara has three interlinked branches: a documentation branch, which is in charge of the registry; a normative branch, which includes an accreditation committee aimed at ensuring the accuracy of the data provided to the registry and promoting better quality of service to patients; and the academic branch, which organises annual meetings and offers continuous education programmes (called PECOnline). Many of RedLARA's activities, including the registry and the professional academic activities, have been funded through financial support from pharmaceutical companies.[12]

One of the first tasks accomplished by the newly formed network was the formulation of a consensus on the ethical and legal aspects of the use of ARTs (Zegers-Hochschild, 1998). Forty-two clinics from eleven countries participated. Mexico was represented by Gutierrez Najar and Silvia Allende (a physician who briefly worked with Gutierrez Najar). The document covers five areas: the marital status required for eligibility, gamete donation, embryo cryopreservation, embryo research, PGD. These issues were selected due to the relevance they had for the legislation process in the region. Previous to the consensus, the network conducted a study focusing on the attitudes among AR users in Latin America; however, the study only covered three countries, Chile, Colombia and Brazil (Zegers-Hochschild, Galdames, & Balmaceda, 1999), hence it only represents views from those countries. Nevertheless, the conservative agenda of RedLara's consensus is not far from the conservative nature of the national consensus presented in 2012 (Kably Ambe et al., 2012); both seek to respect the traditional heterosexual view of the family.

RedLARA's consensus stipulates that assisted reproduction is acceptable for established heterosexual couples (an official marriage certificate is not indispensable) and explicitly rejected the use of these procedures for single women. They accept gamete donation as long as the woman is below the average menopause age and after the recipient couple has been evaluated by a professional, in order to assess their emotional

---

[12]Merck, Serono, Organon have been the most prominent throughout the years.

suitability. They suggest gamete donation should be done fully anonymous, keeping all information confidential, and with no possibility of linking gamete donors with recipients or with the child. In 1995, PDG was still considered an experimental procedure so they suggested treating it as such. They only accept research procedures that do not affect the viability of the embryo to become a baby. Finally, concerning cryopreservation, they suggest linking it to an adoption programme and to try to freeze as few embryos as possible. As mentioned above, both the RedLARA and the national consensus are very conservative, and by being so, they help keep ARTs within the realm of therapeutic procedures seeking to restore a function that was biologically given to the species: sexual reproduction.[13] Uses of ARTs that go beyond this purpose, uses that allow for biologically impossible parenthoods, that are subversive to the dominant cultural mode, are rejected (Hables Gray, 2002); for example, ARTs used by single individuals or same-sex couples who want to become parents. However, in Mexico there are clinics, certified by RedLara, who not only accept single individuals and same-sex couples, but in fact tailor to them specifically.

As of 1997, RedLARA requires clinics to undergo an accreditation process, if they want to be part of the network and report their data to the registry. This process is structured following the peer review process used in academic journals and many research funding bodies. It is a voluntary process that is time consuming, costly, and with no direct advantages in terms of the national regulatory level. In order to be a member of this network, clinics have to request to become part of the network, present their results as requested by the registry, pay the accreditation visit fee, and receive the accreditors who will come to visit the centre.[14]

---

[13]Chris Hable Gray's categorisation of different sorts of cyborgs is a useful tool to understand the conservative nature of these documents. He states that the integration of technology into what used to be natural systems can have different purposes, "cyborgs can be restorative, normalizing, reconfiguring, and enhancing" (2002: 11). Along this line, these consensuses accept the use of ARTs when it is a restorative or normalizing use (restore reproductive function or promote 'normal' reproduction), not when they are used to reconfigure or enhance reproductive practices (promote new family formations, select certain characteristics, etc.).

[14]The visit is done by two accreditors (usually a clinician and an embryologist), who must themselves belong to an accredited centre and have proper training to do this accreditation process, there should be no conflict of interest (i.e. not have any relationship with the centre they will visit), and ensure they will respect the confidentiality of the information received during the visit.

In addition to this mostly administrative steps, they need to meet other requirements as well: (i) document that they "are legally constituted as a clinic" After all, it is the clinic, and not the people who make up the clinic, what receives the accreditation. Requesting the paperwork to prove that the clinic is a legal person is another way in which the AR clinic is distinguished from the OBGYN office. (ii) They need to comply with the health conditions of their country and the ones requested by the network. The local health requirements may vary between countries since some have more regulation than others. (iii) They have to prove that they have the material infrastructure and knowledge to perform ICSI and to cryopreserve embryos. (iv) Their success rates have to be equal to or above the average published by the network, their multi-gestation rate cannot be over the limits stipulated by the network (RedLara document: Norms for the accreditation of AR centres and their andrology and embryology labs, 2011). All these requirements and standards limit the possibilities of who is part of the network. Therefore, the data they compile is not representative of the country or the region. It is simply a snapshot of RedLara approved clinics; in other words, clinics who share RedLARA's views and values.

As mentioned above, RedLARA certifies the laboratory, not the individuals. During the visit, the accreditors will evaluate the laboratory in terms of its environmental conditions, its equipment, its control systems, its operation manuals, and the record keeping strategies (the network produces manuals and standards that clinics must comply with). They will also verify the validity and consistency of the data reported to the registry, the way patient records and lab records are produced and kept, consent forms and the process of obtaining consent, and the staffs' credentials. This evaluation process is computed and presented before the accreditation committee, who will grant the agreed status according to the majority of votes (the status can be: accreditation, conditional accreditation, association, conditional association, and rejected).

While there are many Mexican physicians and embryologists who regard the network as a productive organisation, valuable in its objectives and accomplishments, and worth being part of, there are also those who are critical of the network. During the 2017 annual meeting of the AMMR, at the session where they were discussing RedLARA's registry, several objections to RedLara were made. For example, some were concerned about who can access the information, what can be done with it, and the purpose of this documenting endeavour. Others had concerns because the network has always been overlooked by the same person, Zegers-Hochschild. Several complained that the network had accredited a specific clinic who was well known for its marketing strategy which guaranteed, in their TV, radio, and press advertisements, between 90 and 100% success rates, a practice many saw as deceiving to the patients and unfair to fellow clinics.

The amount of work and the many requirements that are needed in order to be part of the network puts many clinics off the process and others simply get rejected. Of the estimated 100 active clinics in Mexico in 2018, RedLARA's website reports only 38 clinics as accredited and two as affiliated[15] (RedLara's website consulted 26 December 2018). Therefore more than half the clinics (i.e. procedures, births, patients, embryos, etc.) are not represented in the data produced, consumed, and circulated by RedLARA. This situation is similar in the rest of Latin America. Hence, if we aim at using the information provided by RLA for statistical analysis, the information that the register produces is skewed due to the selection bias previously noted. However, if we read this information as a narrative, it proves quite revealing. As they themselves state: "the data collected by the Registry are a source of analysis, not only to quantify the number of procedures performed in Latin America, but also as a way to detect the changes in the medical conduct, to implement education courses, and understand the evolution of the technologies in relation to the characteristics of our environment" (2001). In the following section, I look at these registries as narratives that can tell us a lot about the field and as instruments that promote a particular narrative of the field.

---

[15]To be accredited clinics need to comply with more requirements than for affiliation.

Documenting as Narrative

Throughout the years, the registry has changed in terms of the type of information they require clinics to provide, how they ask them to do so, and the way the registry reports this information in their compilations. These differences can be read as changes in the narrative that the professional community is telling about assisted reproduction, about reproduction as biological and now technological phenomena, and about their field. These narratives show what their concerns have been, the sort of technologies they have used, what they have considered worth monitoring, and what not. The overall plot of these reports is how ARTs have disseminated across Latin America and how, a very specific group of clinics in this region, have adopted these technologies.

Read historically, comparing the reports throughout the years, it is possible to trace when certain procedures or practices were introduced, maybe suggesting they became important enough to be reported, and when others were abandoned, maybe suggesting their importance decreased; for example, the decline of GIFT and the raise of ICSI. In 1994, only two years after ICSI was developed, it was already being used more often that GIF; and by 2000, GIFT had mostly disappeared and ICSI was practised more often than simple IVF. The use of artificial insemination only started to be registered in 2007, even though it is a procedure available since the 1950s (at least). Fertility preservation began to be reported on in 2012 and, although the reports suggest that assisted hatching has little effect on pregnancy rates, its use was still reported in 2014.

Like the overall procedures, the different steps of the process (e.g. male and female diagnosis, ovarian stimulation protocols, birth process, etc.) are also articulated in particular ways, producing thus specific narratives. Some of these steps are included, implying they are considered relevant for the outcome, while others are excluded, possibly for the opposite reason. This is particularly clear in the tables were they correlate variables (e.g. "age distribution or woman undergoing IVF/ICSI" or "delivery rate per embryo transfer according to age of woman", both in Zegers-Hochschild, Schwarze, Crosby, Musri, & Souza, 2013).

The selection of which variables were correlated suggests that certain relations are worth thinking about, while other relations are left hidden, more difficult to see, or completely omitted.

Up until 2010, every report included a table on the couple's diagnosis with five categories: tubal factor, male factor, inexplicable infertility, multiple factors, and other female factors. After this date, this information was no longer included. Why did they stop reporting this? Is this information no longer important? For decades physicians would repeat that it was of paramount importance to have a proper diagnosis before starting any sort of treatment, because it was the diagnosis what determined the procedure to be used. However, not including the diagnosis in these reports seems to eco Serviere Zaragoza in *Medicina Reproductiva* where he points out how the importance of diagnosis has diminished "[…] the selection of the technique responds more to the results and complications of the technique, than to the diagnosis" (2003: 289).

Information concerning woman's health is also not present. While her age is always mentioned, and up until 2010, also was the general diagnosis of the infertility, they omit if it is a case of primary infertility (never had a pregnancy before) or secondary infertility (existing prior pregnancies), as well as other health issues which frequently are associated with infertility (e.g. diabetes, overweight, thyroid problems, metabolic problems) (Abalovich et al., 2007; Cardozo, Pavone, & Hirshfeld-Cytron, 2011; Mulholland, Mallidis, Agbaje, & McClure, 2011; Raju et al., 2012; Zander-Fox, Henshaw, Hamilton, & Lane, 2012). Information regarding the ovarian stimulation protocols is now absent, but when they were reported, they were never related to diagnosis (through a table indicating outcomes for example), were these not considered related? How can we consider ARTs (AI, IVF, ICSI) therapeutic if they are not related to a particular diagnosis and if the stimulation protocol does not relate to the diagnosis?

If the overall health state of women is not reported, the information concerning men is even less so. The registry focuses on infertility and the technologies from a gynaecological perspective overlooking the andrological aspect. The reports have never mentioned men's age, the evaluation and diagnosis of their sperm sample, the way the sample was obtained (via masturbation or surgical procedure) or the treatment they received (if they did). This omission is done in the context of a repeated claim that male factor infertility represents at least 30% of the cases (Red LARA, 1996, 1999, 2000, 2001, 2002, 2007). Does this suggest that the sperm has no role in the fertilisation, development of embryo in vitro, implantation, and development in uterus? The only moment in which sperm donation appears is in the very short section on AI, which was included as of 2007.

When reporting the outcomes, they focus on the number of clinical, ectopic, and multiple pregnancies, the number of live births, the week when birth occurred (whether they are premature or not), the cases of malformation, and the number of abortions. However, they do not report the number of c-sections or induced labour. Furthermore, there is no registry of longitudinal follow-ups, neither of the woman nor of the child. And there is no information about surrogacy cycles, which is absent even from the evaluation for the accreditation process.

In contrast to the issues that have been omitted, there have been others that have been overemphasised. For example, for nearly two decades multiple embryo transfer has been a practice causing great concern because it frequently results in multi-gestation. In 2002, it was declared an epidemic in Latin America (RLA, 2002). In most of the reports, they offer information regarding the frequency of multiple-gestation, the order, the gestational age when they are born, their weight. This issue is of such concern to the network, that it is a point that weighs heavily in the evaluation of the clinic when undergoing the accreditation process.

These reports also offer a narrative of the region. These reports tend to order countries in terms of how many clinics report to the RLA and how many cycles they initiate annually. Under this criterion, Mexico has been among the top three countries (together with Brazil and Argentina) almost ever since the registry began. However, the order changes if the criterion is what RedLara calls availability, which is the total number of cycles divided by its total population, a data presented as the number of cycles per 1000 inhabitants. In this last case, Mexico goes from being among the top three to being the eight (in 2014)[16]; under Argentina,

---

[16]Argentina (348.8), Chile (211.0), Uruguay (150.3), Brazil (133.4), Peru (108.1), Panama (97.7), Ecuador (81.6), Mexico (76.6). (Zegers-Hochschild, Schwarze, Crosby, Musri, & Urbina, 2017).

Chile, Uruguay, Brazil, Peru, Panama, and Ecuador. However, in order to make some sense of these numbers it would be useful to consider, for example, the age distribution of the population –some of these countries (e.g. Chile, Uruguay, and Paraguay) have an older population than Mexico. Also relevant is that Argentina, and now Uruguay, are the only two countries with universal access to assisted reproduction as part of their reproductive rights (Zegers-Hochschild, Schwarze, Crosby, Musri, & Urbina, 2017). Finally, it would be important to consider that the number of cycles reported to the registry is not the total number of cycles initiated per country, since only some clinics are part of the network.

Who is the target audience these reports have in mind? What is their purpose? Registries can have many uses: (a) to inform users, (b) to compare the performance of clinics or countries, (c) to compare an individual clinic's practice through time, (d) to compare procedures and assess their potential for success, (e) to establish epidemiological data/analysis, etc. Neither of these uses is really possible with this registry because of how these reports are narrated. Although we can know how many and which clinics participated in the registry (since there is always a list of names at the end of the report) and we can learn about the number of cycles and procedures used overall in each country (since the numbers of the individual clinics are added to offer a country's result), we cannot know how many pregnancies were achieved, how many of these were singleton or multiple, nor how many resulted in lives births in each country (let alone per clinic). Therefore, patients, who can access these reports because they are online and open-access, cannot use the information to make an informed decision about which clinic and which treatment to use. Scholars cannot use this information to assess how success is distributed across countries, nor to carry out an epidemiological analysis. Only clinics can use the data to, roughly, compare their practice and outcomes to the overall regional practice and outcomes. However, how much can they learn from this comparison given the shortcomings of the information presented above?

Documenting data is a complex task, particularly when the task is to register the outcomes of a technology that is constantly changing, and when there are all sort of things people are interested in keeping track of, as well as many things they want to be kept off record. What and how information is collected, computed, and narrated, contributes to the narrative the registry offers, as well as the narratives that can be constructed from this registry. One of the few things we can read from this registry is that there has been a steady increase in the number of Mexican clinics interested and able to participate in the network and in the registry. This is, partially, due to an increase in the total number of clinics active in the country. This increase is the subject of the following section.

Having said all this, why do I consider the RedLara important to the Mexican AR system? The network embodies a particular perspective of how medicine and science should be practised. Although this perspective might not be shared by all members of the networks or by the members of the AR community that are not part of the network, it does have an impact on the region's AR practice. On the one hand, the network is strongly influenced by its founder, and on the other, national AR systems are now turning to the RedLARA for help in creating their national registries (AMMR annual Conference 2017). This perspective looks towards the anglo-sphere for comparison and guidance regarding how to practice assisted reproduction, both clinically and scientifically. Zegers-Hochschild, for example, considers that there are several barriers to conducting clinical research in reproductive medicine in Latin America. These barriers, he explains, are the result of what he sees as Latin America's relationship to his view of science: "Societies in Latin America are not scientifically driven and therefore, the allocation of human and economic resources to research is meagre, as a reflection of this as well as other cultural and economic realities" (Zegers-Hochschild, 2011: 802). Furthermore, he considers that in Latin America "decisions are triggered by emotional and religious influence rather than using the evidence provided by science and technology" (Zegers-Hochschild, 2011: 802). During AMMR's annual meeting mentioned above (in 2017), both Zegers-Hochschild and a representative of the SART was there to push the agenda for the need of a national and regional registry. The Mexican physicians there were wondering if the best option was to participate in the RedLARA's registry or to work towards a national, official, and compulsory one. If so, who should coordinate it, the Health Ministry, by COFEPRIS, or by the AMMR?

RedLARA's registry mixes an evaluation process and a registration process. While this works as a way to standardise practice (e.g. lexicon, definitions, and procedures) giving the field strength, it also sets limits to what can be registered and thus what can be done with the information it compiles. In sum, RedLARA is working towards becoming an international epistemic community capable of claiming epistemic jurisdiction in the field of assisted reproduction and thus, of mediating between the medical community and the local governments (Quark, 2019).

Increasing Availability

Simultaneous to the process of diffusion is the process of acceptance. If an idea, theory, or technology is not accepted and adopted it will eventually fade away. Hence, in order for an idea, theory, or technology to be available it needs to be accepted by those who are going to offer it. Therefore, the fact that assisted reproduction appears in INPers "Black Book" suggests it had been accepted and adopted as a valid and important field. "The Black Book" was a celebratory publication, produced by Samuel Karchmer, the hospital's director, compiling the activities and achievements of the different services of the hospital since it had been given the status of National Institute in 1993. The librarian at INPer's one-room library lent me the only copy of this book, "you will find much of what you are looking for there" she said, and she was right. This book was more than celebratory, it worked as a testimony of what INPer had done in the past ten years, a way of justifying its new title of National Institute, and of accounting for the public investment it has received. It was also a site where emerging fields were being consolidated.

Each of the services offered at the institution was assigned a page with the picture of its director, a small explanation of what the service offered, and the results it had achieved. There were three services caring for fertility problems, the Andrology service, the Sterility service, and the AR service. The Andrology service was focused mainly on developing standardised procedures to process, study, and report results on sperm samples. No images accompanied this section. The Sterility service received patients, carried out all the diagnosing procedures, and treated patients with the surgical procedures. The image presented in the Sterility service's page was precisely two doctors performing a surgical procedure. If the treatments offered there were not enough, patients were sent to the AR service, where they offered high complexity ARTs. Between 1989 and 1993, the service had performed 1627 procedures and achieved 212 pregnancies via AI, 40 using GIFT, and 9 with IVF. The images illustrating the section depicted, on the one hand, the head of the service, the white coat physician Kably Ambe sitting at his desk, and the embryologist, a man wearing scrubs in the lab sitting at a laminar flow hood and looking through a microscope. These images show how this field was beginning to be recognised as having two equally important areas: the medical (the physician) and the biological (the embryologist). The visibilisation of the different areas has progressively increased. Now (in 2019), clinics commonly include pictures of the nurses, administrative staff, and support areas (like psychology or nutrition) in their websites, and the AMMR has recently included courses for nurses in the annual meeting and in their monthly activities.

For almost five years (between 1985 and 1990), apparently no new clinic opened, there were only two cities with clinics, Mexico City (INPer and AGN) and Monterrey (IECH). But between 1990 and 1999 the number of AR clinics multiplied fivefold: from 2 to 17, these were located across eleven cities.[17] This is a brief story of how this increase in availability of clinics could have happened.

Between 1988 and 1990 AGN performed 91 IVF cycles and 52 GIFT cycles, but their pregnancy rates were low, 5.48% with IVF and 36.7% with GIFT. This led them to rethink their methodologies and implement significant changes (Gutierrez Najar et al., 1992 (6)S1: 100). In 1991, after these changes were implemented, they achieved their first successful IVF. With their first successful IVF, a slow but growing demand for these services was being felt. The city was also becoming crowded, making the clinic in the south difficult to access for people in

---

[17]Tijuana, Ciudad Juarez, Estado de México, Guadalajara, Monterrey, Puebla, SLP, Hermosillo, Tamaulipas, Veracruz, and Mexico City.

the north of the city. With this in mind, Gutierrez Najar got together with a group of specialists (around 1993) and opened a second health clinic, a larger healthcare facility focused on women's health. This place was called CLASA (Clínica Lomas Anahuac S.A.), it was located in the relatively new area called Interlomas.[18] At this point the only private clinics were the one in Monterrey (IECH) and the one in Angeles Pedregal (AGN); a year later, in 1994, Kably left InPer and established his private clinic in Palmas. However, not far from CLASA Vázquez Raña was building his second Hospital Angeles, where they were thinking of having a second AR clinic; he invited Kably and his team to transfer their clinic to the hospital (this was in 1998–1999).

Circa 1993, "two younger physicians who had studied in IVI-Spain returned to Mexico with new technologies, a new outlook on reproduction, and offering these services at lower cost" (AR Specialists C, Interview, 2007). These two doctors, Gutierrez Gutierrez and Gallardo, opened their clinic in Leon, Guanajuato. Other physicians and embryologists also went to IVI, "there is an entire generation of embryologists who studied at IVI" (Embryologist 0, Interview, 2018) and the influence of the Spanish clinic's modus operandi soon become evident. When this happened, IVI was not really that well known. In fact, "some say these physicians helped make a name for IVI" (Embryologist C, Interview, 2007). This same embryologist also felt that the peculiarity of IVI was their understanding of what happens in the lab: "They recognise that the biologist and the lab are central, crucial for good reproductive rates. So they invested heavily in training their biologists and improving their lab procedures" (Embryologist C, Interview, 2007).

Who is IVI and why are they important in this story? IVI is the Instituto Valenciano de Infertilidad, established in 1990 by two Spanish gynaecologists, Antonio Pellicer and José Remohí (both with training in the USA). From the onset their interest was both the clinical aspect of AR and the research aspect; so, in 1997, they established a research

---

[18]Interlomas is a wealthy suburb located in the north-west part of the city. At that time it would take about an hour to get from Angeles Pedregal to Interlomas, so this second clinic would be serving another population.

foundation together with Carlso Simón. Five years after they had opened their first clinic, they began expanding, first within Spain and then, in 2002, they went beyond their borders and opened their first international clinic in Mexico. Before opening their branch in Mexico City, they studied its viability and concluded it was a profitable market (IVI, Information Session, 2007). A physician once told me that "IVI in Spain is probably the strongest centre, with top researchers, but opening a branch in Mexico was an economic decision, they were not here to teach or to contribute to the scientific quality of research in Mexico. They were here to do business" (Physician K, Interview, 2017). This business orientation was evident in their effort to branch out. Soon after opening a branch in Mexico they began to expand to Portugal, Chile, Panama, Argentina, Brazil, EAU, Italy, and the UK. In 2017 IVI merged with RMANJ (Reproductive Medicine Associate of New Jersey) making IVI-RMANJ an international reproductive business with over 65 clinics in 11 countries: "being in the United States is a dream come true and RMA of New Jersey proved to be the best companion. They are commercial and scientific leaders in reproductive medicine in the USA. This union will allow us to enhance our field of investigation, the fundamental pillar of IVI since its inception which has enabled us to achieve high rates of success – a fact that our patients value when choosing our services" (José Remohí[19]). In 2009 they established a reproductive genetics analysis company, Igenomix,[20] also in collaboration with Carlos Simón.

Throughout their nearly twenty years of history, they have not only opened clinics all over the worlds, developed techniques and procedures, organised conferences, and trained many physicians and embryologist, they have also established a way of building, managing, and practising the business of human reproduction. Today, the narrative of their reproductive business is told through apps, audits,

---

[19]José Remohí a founding partner of IVI and member of the board of directors. https://www. rmanj.com/about-us/ivi-rma-global/, accessed 29 January 2019. In addition to the clinics.

[20]A genetic testing company that specialises in reproductive genetics analysing the genetic composition of ova, sperm, embryos, and endometrium.

international partnerships, and a transnational presence. In 2018, their website invited you to download an app that will "offer you a complete service throughout your treatment", they displayed a banner declaring they were the "largest group of assisted reproduction in the world, in partnership with RMA they have become IVIRM Global with 65 clinics in 11 countries", and they claim that their success rates are audited, "9 out of 10 couples will achieve their objective" (visited in December 2018).

However, as a clinic, IVI in Mexico faced a more difficult scenario. When I first visited their clinic in 2007, they were in a building located in a residential area within small commercial enclave with office buildings and shopping malls. A few years later, they moved the south part of the city into a new and much larger building, with a modern high-tech design. This move would have suggested that they had been successful, had grown, and thus needed more space. However, by 2016 they had left those big modern offices and merged with a Mexican clinic, moving into their premises. Why were they not so successful?

One embryologist suggested that the problem was that "IVI Spain never took into account that Mexico and Spain are two distinct places with very different cultures, requiring different bedside manners. Mexicans like being seen by the same doctor all through their procedure and their pregnancy. They don't like being a number. They wanted the Mexican clinic to work as the Spanish one, but this was not possible" (Embryologist C, Interview, 2007). Another embryologist believed it was the way they managed patients: "It is not the Spanish style, it is the IVI style [...] Technically it is the same thing everywhere, one might use a square box and the other a round one [...] but what makes the difference is the way they handled patients" (Embryologist K, Interview, 2017). Others believed it was the way they managed their team "At IVI the administration is one thing and the medical area is another [...] They never considered the way Mexicans operate, the contract didn't consider this. The Mexican doctors saw this was a business, they observed, learned and then went and set up their own clinic. There came a point in which the physicians working there said 'I can put up my own clinic' And they started to leave and put up clinics all over the city taking with them embryologists. They started buying people out and taking them with them" (Embryologist O, Interview, 2018, a similar idea was expressed by Distributer F, Interview 2017). In sum, cultural differences, particularly in terms of the patient-physician relationship and the working environment, made IVI less of a success story than what they expected when they opened in 2002. However, certain aspects of their model, particularly the commercial aspects, permeated to the different clinics that spun off IVI. For example, the practice of promoting extremely high success rates or of guaranteeing a baby.

The discussion about IVI also underlines the place embryologists were gaining within the clinic. While the first experiments with IVF conducted in the first three clinics were done by a team of male physicians (Martin, Gutierrez Najar, and Lichtenberg; Santos Halisack, Hernandez Ayup, and Pedro Galache; Kably and Reyes Fuentes) within less than two years these teams were incorporating other professions to take over the lab-work. This was when veterinarians, biologists, and chemist made their way into this gynaecology space and became the laboratory staff; becoming thus embryologists, the ones in charge of the management of the gamete and embryo labs as well as of the procedures that take place there. However, IVI did not only introduce new ways of handling gametes and embryos in the lab and administrating the clinic, it also introduced a new way of viewing the *business* of reproduction. They came to Mexico after performing a *market study* to see how *profitable* it would be and they saw there was a *business* opportunity. Although the clinic as such did not prosper as they probably expected, their way of thinking about assisted reproduction, their way of practising it, and the constitution of the place where it is practised did influence the Mexican AR system. Since they began receiving Mexican AR professionals (physicians and embryologists) in the mid to late 1990s, then after establishing their own clinic in 2002, and in 2008 signing an agreement with UNAM so the students could do internships at their clinic, there have been several spin-offs of IVI, all following a marketisation and economisation logic. In the following section, I trace the development of the AR clinic during the first ten years of the twenty-first century.

The AR Clinic

> [...] a centre that offers assisted reproduction should have a series of human and infrastructural elements that allow it to offer acceptable results, as well as a guarantee that they will behave ethically and professionally. (Tlapanco Barba, Medicina Reproductiva, 2003: 293)

While the protocols of low complexity ARTs (e.g. ovarian stimulation, programmed coitus, and even some types of AI) used in the 1980s could be offered in a standard OBGYN office, the new protocols and the high complexity procedures (IVF and ICSI) demand highly specialised equipment, tailored facilities, specific knowledge and technique, and multidisciplinary teamwork. The standard gynaecological office, with only one doctor and a nurse, was no longer enough. Therefore the introduction of high complexity ARTs meant a transformation of the gynaecologist's workplace into an AR clinic, it meant the birth of the embryo lab, the emergence of the embryologist, the reconfiguration of the physician, the introduction of the administrative staff, and (more recently) the configuration of the AR nurse (González-Santos, 2011).

Financing the costs of the specialised equipment and the qualified staff is difficult for someone who is just starting a practice or lacks the clientele to cover the expenses. This dilemma was seen as a problem by some and as an opportunity by others. During the last decade of the twentieth century and the first decade of the twenty-first century, four models and strategies to deal with the problem of how to finance opening an AR clinic were tried out. In the referral or outsourcing model, physicians without the skills or tools to perform the procedures, but with a patient who needs them, would refer their patient to a clinic that does have the skills and tools hoping that, upon success (i.e. preganancy), the patient would come back to them to look after their pregnancy and other gynaecological issues. Others would perform the procedures themselves, but sublet the equipment and staff; in other words, patient and physician would go to a third party's clinic to use the equipment and staff. The owner of the equipment would benefit from the extra income to pay for the laboratory's maintenance. A couple of doctors became consultants, providing gynaecologists the necessary AR equipment and offering them consulting services and guidance throughout the process. This "consulting model" was tailored to the established gynaecologists who while having demand for it, did not have the necessary infrastructure or experience to offer assisted reproduction. By subscribing to this model, non-AR gynaecologist could have access to an AR clinic, with all the facilities, specialised advice and professional high-tech equipment, allowing them to offer AR services to their patients without having to give them up. A fourth option was the "network model". Clinics that have followed this model have one large, well-equipped hub clinic, which centralises the expensive aspects of the protocols, and many smaller satellite clinics across a large geographical area that offer the less technologically demanding parts of the protocols. By doing so, the network can reach out to more corners of the city and country and thus attract more clients. One example of this is the RMA Clinic, where Benjamin Sandler works, the main office is located in New York and the Mexican branch is located in Mexico City. In this case, the first part of the chemical part of the treatment is offered in the Mexican branch (diagnostic tests, preparation, and ovarian stimulation), after which patients fly to the clinic in New York for the surgical part of the procedure (egg aspiration, fertilisation and embryo transfer); eggs, sperm, and embryos are handled in New York. One last model was the franchise, one clinic sells the franchise to new doctors.

These different models of private clinics (outsourcing, subletting, consultant, network, and franchise) work in different ways. While in some cases the physicians are the owner and the ones who have invested in the equipment and staff, in other cases there are investors and the physicians are either partners or employees, and yet in other cases the owners of the clinics (be them investors or physicians) buy a set of procedures and know-how (franchise). How clinics are configured and managed matters. Proper management of the staff, the equipment and supplies, the finances, and the patients is crucial for success, understanding success as both the achievement of pregnancies and having a profitable sustainable clinic.

Management of a clinic influences what sort of procedures and techniques can be performed and the number of patients a clinic can care for at a given time. IVF labs can handle a limited number of patients per cycle and can only offer a selected sort of procedures. These limits are shaped by the infrastructural composition of the IVF labs. The maximum number of patients they can manage per cycle depends on, for example, the number of incubators, nitrogen tanks, microscopes, and ICSI benches they have, the number of embryologists on site, and the way these (humans and machines) work together. While the type of services they can offer depends on the type of tools they have and the access they have to certain products. This has been a point frequently brought up by embryologists and physicians during conferences:

the type of drug we can use depends on how accessible they are in terms of disponibility, costs, and security. (Physician, AMMR, 2007)

[…] Labs that are working with one single embryologist and have high number of cases, they need a medium they do not need to change […] The time-lapse gives you a lot of information, but it requires a full time embryologist dedicated to analysing the data. (Embryologist workshop at ISMAAR, 2015)

Regardless of which model physicians follow to care for their patient, and AR clinic has to have certain basic things to be able to offer a service that is ethical, professional, and that gives patients the possibility of success. These things include, for example, high-tech equipment for conducting the procedures (such as micromanipulation systems, ovum aspiration pumps, CO_2 incubators, etc.) and environmental control systems and methods (such as special paint on the walls, double filtered air, controlled temperature, controlled lighting, pressurising modules, and laminar flow chambers). Likewise, people working at AR clinics— be these physicians, embryologist, or technicians working at the labs— are expected to fulfil certain certification and membership requirements (i.e. diplomas and certifications issued by validated institutions and membership to a recognised AR professional association).

Although these elements contributed to the characterisation of what an AR clinic should be, there are three other elements that clearly distinguish it from other biomedical consultation offices: the way they are named, their digital persona, the use of marketing schemes, and their financing options. Clinics usually have a name, a slogan, a logo, an increasingly more, a digital persona (the combination of their website, their Facebook page, their Instagram and Twitter account, and some times, even a YouTube channel) through which the clinic (re)produces its own identity, independent from doctors. Clinics are named using acronyms which highlight specific things and by doing so, they construct the field of assisted reproduction in particular ways. Sometimes they stress the female role (e.g. Inmater), the aspect of life (e.g. Instituto Vida), or they prioritise conception (e.g. Concibe) or genetics (e.g. Ingenes). The name, logo, and the clinic's digital persona, facilitates a process of individuation where the clinic becomes independent from the physician, allowing both to exist without each other. There have been cases where the doctor who founded and directed a particular clinic has to leave it, yet the clinic continues to exist with other physicians directing it, while the founder establishes a new one. This individuation process has changed the way patients relate to and talk about the clinics they seek treatment at. Users now mention the clinic by its name, no longer by their doctor's name. Consequently, due to the structure the new clinics are taking, the patient is no longer "the doctor's patient", they are becoming "the clinic's patients". Clinicians have found that this offers them the possibility of working as a team, which is a practical way of dealing with the protocols' demand for constant supervision. In spite of patient's criticism to this way of practising medicine, within the realm of AR it is becoming a standard practice.

Embryologist and Their Laboratories

> The lab is probably the centre of where the transcendental changes have occurred. […] The use of sequential medium has lead to changes in the transfer date […] ICSI, assisted hatching, PGDs […] more clinics are doing more micromanipulations, particularly the most busy one, you can no longer risk having problems with fertilisation. (Kably, Field Notes, 2008)

The job of an embryologist requires a very particular set of skills. Embryologists perform and manage the gamete and embryo lab activities, they administer the supplies, keep track of the maintenance of the machines, monitor the environmental conditions as well as the development of the embryos, they evaluate and prepare the embryos for transfer, and process the gametes for fertilisation. In occasions, they also have to produce their own tools. In the past, for example, embryologists would have to make their own medium (a problem in places where the water quality was not right for the mixture[21]), then they had to make their own ICSI needles (I was shown how to do this, by warming a glass straw and then braking it). Embryologists many times had to resort to using all sorts of improvisations to resolve everyday problems; I have even seen how they use mundane objects (like tennis balls) to stabilise ICSI tables, for example. The thing is, as stated in the opening quote, labs have undergone many the changes, and to keep up with these embryologists need training.

The problem with training is where to go get it? The most common practice is to go to other labs and be taught by other embryologists: "Biologists can only receive training in other AR clinics, so these clinics have to accept training biologist who will later on go to another clinic that might end up being competition for them. So sometimes, the training clinic only gives their students half the information, so they can't steal the techniques and know-how and then become competition. This was a way that the boycott was done. If one clinic was worried about the competition, they would not allow new biologists to come to their clinic" (Embryologist C, Interview 2008). In the early 2000, they would also go to IVI. As Internet became more accessible and viable, embryologists could now read journals and use the Internet as a training tool also. Things also changed within the AMMR, while in the 1980s and early 1990s conferences were mainly organised by and for physicians focusing on the clinical aspect of assisted reproduction, in 1997 the first pre-congress course for embryologists was programmed (García Villafana, 2018). At these events, I witnessed great degree of teamwork. For example, at the 2017 annual conference of the AMMR the pre-congress course, organised by embryologists for embryologist and where

---

[21]One of the first difficulties they faced at the beginning of the IVF program was the water quality, Monterrey does not have the type of water that favors the procedures, it was not adequate for the medium which, at that time, was house made. When HAM F10 became available, a premade medium, their first success soon followed.

only embryologists attended (although open to all conference attendees), they discussed the use of a new sperm sorting tool called micro-
fluidics. After a couple of oral presentations, they all gathered around
microscopes to experiment hands-on experimentation with these new
tools. Likewise, at the International Society for Mild Approaches to
Assisted Reproduction, (ISMAAR) conference in 2015 for example,
talks were introduced or followed by a questionnaire that was digitally
answered so the answers were computed and presented as percentages
immediately. This allowed for a discussion about how they do things
differently and why, sharing their experiences, problems, solutions, etc.
There was one more important actor at these events: representatives of
the import and distributing companies.

Importing and properly distributing equipment and supplies can
be very complicated. As one embryologist explained, "if you want to
import your own equipment, it will take you six months and it will be
about 25% more expensive, due to all the money you need to give to
'lubricate the system' but if you use them it will take only three months
and cost less". "Them" was any of the handful of companies that saw
in this import-distribution problem a niche for a business opportunity.
The oldest of these Mexican companies was Demesa, located in Mexico
City; it has been importing equipment and supplies for the areas of
urology, laparoscopy, and assisted reproduction for over 35 years. The
newest is Nanogbiotec, established in 2016 by four young embryologists, two brothers and a couple, all formerly working at a clinic in
Guadalajara. Their experience as clients of these companies led them
to consider putting up their own company. Probably the most spoken of is Lepsi Prisma.[22] This import and distribution company from
Monterrey was founded over 15 years ago and was directed by Miguel
Santos Halisack, brother of the doctor Roberto Santos Haliscak from
the IECH. "They have a good chunk of the market because they sell at
a better price and because the Engineer Michael Santos was great, he

---

[22]They distribute equipment like microscopes, incubators, and laminal flow bells, as well as supplies such as cultivating medium, syringes, micro-pipets, and other lab material. They also offer
maintenance services, sending technicians to clinics to check and repair the equipment.

was a fantastic person who knew how to win people over. He would invite the embryologists to the AMMR conferences" (Embryologist O, Interview, 2018). One last company that made presence at these events was Corne Fertilidad; however, they only distribute gonadotropins and FSH, both drugs for patients.

The changes within the lab and the importance of the embryologist are also evident at the conferences, particularly at the conference exhibit hall. Claudia, a woman who has been in the business for well over a decade, working both at clinics and distributing companies, confirmed what I had suspected "each year there are more and more exhibits" (2017). In 2007, when I went to my first AMMR annual meeting, the exhibit hall had no more than just a few stalls selling ultrasounds and books. In 2011, at a curso de actualización organised by COMEGO, there were only four stalls and only two were promoting distribution companies. Then in 2015, at the joint conference organised by ISMAAR and the IECH, two of these companies (Lepsi Prisma and Corne) participated in the course organised for embryologist. But the big changed was evident in 2017 at the annual meeting of the AMMR, the exhibit area was big with exhibits several feet wide and long with a little over 20 stalls. There were two sperm banks, specialised drugstores, law firms, distribution companies, pharmaceutical firms, a genetic testing company, and the AMMR.[23]

Analysing the exhibit halls of these conferences can offer a very particular view of the development of the field. First, we see an increase in the awareness of the role of the embryologist and the lab in achieving success. We see this not only in the number of stalls directed to the embryologists and their labs, but also in the number of embryologists who were at the exhibit hall. Meandering between the exhibits and interacting with the exhibitors were embryologists, alone or in small groups, asking for information, taking notes, comparing prices and characteristics, and saying thing like "I will bring my boss to see this so

---

[23]These are some of the names of the companies that had stalls at the event: Lepsi Prisma, Nanogbioetc, Demesa, IFA Celtics, Asofarma de México, CENAREM, Besins, Fertifarma, Kitazato, Biolietc, Corne, Ferring Merck, Igenomix, and Gedon Ritcher.

he buys it for me", the boss being the head of the clinic, i.e. a physician. Second, we see the appearance of sperm banks. In this particular event, there were two, one national and one international. Third, we see the emergence of law firms. Overall, the aesthetics that predominates in this exhibit hall is, obviously, a commercial one. In the following chapter, I explore deeper into this process of martketisation.

Summary

In this chapter, I looked at how assisted reproduction was made usable for those who offer these services and how, through the process of making it usable, the universe of the Mexican AR system grew. It expanded in terms of number of practitioners, clinics, procedures, uses, users, etc. This expansion took place in the context of a country that was also expanding in terms of its economic and cultural boarders. In this case, Mexico's expansion was done through the signing of international treaties like the NAFTA, the participation in international cooperation groups like the OECD, and in general, by aligning to the neoliberal project.

During the period this chapter covers, the universe of assisted reproduction expanded. This expansion was fueled by disseminating the knowledge among students and physicians through training programs, conferences, articles, and books. In this chapter, I focused on some of these publications to see how this knowledge was being presented and how the construction of what assisted reproduction is and what it can do has changed.

The expansion triggered and was fueled by the creation of more spaces where these procedures were performed. Throughout the chapter I looked at how new AR clinics began to open in cities other than Mexico City and Monterrey, and how these healthcare sites professionalised and became a very specific techno-scientific site: the AR Clinic. The opening of new clinics also meant the beginning of competitive environment between clinics. This activated a process of marketisation which will be further explored in the following chapter.

The universe of assisted reproduction expanded by the development of more procedures and the flexibilisation of the eligibility criteria. This in turn helped expand the universe of users and uses. Users and uses where moving beyond the confined boarder of "the heterosexual stable couple who has a diagnosable condition for which assisted reproduction is the only option". Because the universe was expanding, practitioners and policy makers intensified their efforts to regulate these services. On the one hand, professionals created new sets of regulations through the production of consensus documents and through certification processes. On the other, the government presented bill after bill to regulate ARTs and installed certification processes to regulate the laboratory activity.

The universe of assisted reproduction expanded by collaborating with people in other geographies. I looked at how Spain influenced Mexico's AR system in terms of how to perform the different ARTs, how to manage a clinic, and how to care for patients and staff; and I looked at how Mexico resisted this influence. I also described how Mexico collaborated with fellow Latin American AR professionals in order to gather force as a community of practitioners in a field that is frequently contested with moral, economic, and medical arguments.

In sum, this chapter looks at the consolidation of the contemporary epistemic constitution of the field (as mostly focused on ARTs), a specific material infrastructure (the AR clinic), a national and international networked community (the linking of AMEE and AMMR and the Latin American network RedLara), and a self imposed normative infrastructure. By doing all this, the field also acquired a more stable sense of specialisation and professionalisation. The multiple assemblages described in this chapter, set the conditions for a new understanding and performing of assisted reproduction and for a process of dissemination through a varied discursive landscape. As I will explore in the following chapter, assisted reproduction in the current century is no longer simply a medical procedure to cure a medically described condition, it is now a tool through which people create people. This tool is part of a consumer's market guided by neoliberal logics and rhetorics. This new way of practising assisted reproduction is changing the way we are made as individuals and as a society.

Detail of a National Portrait

On 3 July 2000, hope hovered over the country like a dense fog that blurs your vision and makes everything look so much more appealing. After 70 years of PRI's regime, in 2000 a different party won the elections, the conservative right wing Partido Acción Nacional (PAN[24]). I argue that PAN's victory set favourable conditions of the marketisation of assisted reproduction (which is the subject of the following chapter). This conservative party held power for two terms, from 2000 to 2006 with Vicente Fox and from 2006 to 2012 with Felipe Calderón. During both terms, family planning and reproductive health were topics that were taken off the table due to ideological and religious reasons (Maldonado, 2017). Instead, the idea of taking advantage of the population bonus[25] was taken up (Welti-Chanes, 2014). The result of this change of perspective was that the goals set by the CONAPO in terms of fertility

---

[24]The PAN (Partido Acción Nacional) was founded in 1939 by Manuel Gómez Morín, a man who stood against liberalism and individualism, together with members of the Jesuit national union of Catholic students (UNEC, Unión Nacional de Estudiantes Católicos) as an opposition to Lázaro Cárdenas. It has been a party that has frequently had to fight internal tensions between its pro-Catholic and pro-secular fractions.

The triumph of the PAN in 2000 was a surprise full of hope, but it was not announced. Although PRI had seldom allowed them to claim the victory, the PAN had already won smaller town elections during the eighties. However, the mixture of years of corruption and abuse of power, the worst financial crisis in the history of the country, and a new actor in the political scene made executing the traditional fraud no longer so easy. The new political actor was the Church, mainly in places like Chihuahua and Oaxaca (but also Chiapas and Morelos). After years of silently watching (and some would say participating in) how the PRI secured and used its power they began to live in their own skin the effects of the economic crisis, encouraging them to brake silence and urge parishioners to vote responsibly, to vote for change, to vote for those who were seeking profound changes for society (de la Torre, 2008). For example, when the 1986 local elections were underway, the archbishop of Chihuahua went even further and declared that not voting and committing fraud was a sin. Then, when a vulgar display of electoral fraud was performed in Chihuahua, the Church openly responded to the extent that the Secretary of State called the Vatican's apostolic delegate to Mexico to ask for his aid in stoping the archbishops' actions (Preston & Dillon, 2004). So, when in 2000 the PAN won the federal presidential elections, PRI could not do much more than let them take office.

[25]The population bonus was a period of time during which there is a possibility to accelerate development due to the composition of the population: the population within working age (between 15 and 65 years of age) is larger than the depending population (younger than 15 and older than 65).

rates, accessibility to and use of contraceptives and family planning services were not met. Furthermore, and in relation to assisted reproduction, it was precisely during these years that AR clinics began to expand and began to resort to advertisements to promote the use of ARTs. The point being that the government was withdrawing messages on family planning from the media, at the same time that messages promoting the use of ARTs were being broadcasted by clinics.

References

Abalovich, M., Mitelberg, L., Allami, C., Gutierrez, S., Alcaraz, G., Otero, P., & Levalle, O. (2007). Subclinical hypothyroidism and thyroid autoimmunity in women with infertility. *Gynecological Endocrinology, 23*(5), 279–283. https://doi.org/10.1080/09513590701259542.

Alvarado Durán. A. (1999). Ciencia y Bioética. Cap. 73 (p. 387). In E. Vázquez Benítez (Ed.), *Medicina Reproductiva en México* (Primera). Mexico: JGH Editores S.A. de C.V.

Álvarez Navarro, M. (1999). La Bioética en Reproducción. In E. Vázquez Benítez (Ed.), *Medicina Reproductiva en México* (Primera). Mexico: JGH Editores S.A. de C.V.

Callon, M., Law, J., & Rip, A. (Eds.). (1986). *Mapping the dynamics of science and technology*. London: Palgrave Macmillan UK. https://doi.org/10.1007/978-1-349-07408-2.

Cardozo, E., Pavone, M. E., & Hirshfeld-Cytron, J. E. (2011). Metabolic syndrome and oocyte quality. *Trends in Endocrinology & Metabolism, 22*(3), 103–109. https://doi.org/10.1016/j.tem.2010.12.002.

Cohen, J., & Alikani, M. (2013). Evidence-based medicine and its application in clinical preimplantation embryology. *Reproductive BioMedicine Online, 27*(5), 547–561. https://doi.org/10.1016/j.rbmo.2013.08.003.

Consejo Nacional de Población. (1999). Introducción. Veinticinco años de planificación familiar en México. *LA situación demográfica de México*. México: Consejo Nacional de Población.

de la Torre, R. (2008). La Iglesia Católica en el México contemporáneo. Resultados de una prueba de contraste entre jerarquía y creyentes. *L'Ordinaire des Amériques, 210,* 27–46. https://doi.org/10.4000/orda.2616.

Fuentes, C. (1993, December 2). Tribuna / TLC, el día siguiente. *El País*. Retrieved from https://elpais.com/diario/1993/12/02/opinion/754786811_850215.html.

Galache Vega, P., Santos Haliscak, R., Batiza, R., Saucedo de la Llata, E., Arenas Montezco, L., García Villafana, G., & Ayup Hernández, S. (2003). Fertilización in Vitro. In E. Vázquez Benítez (Ed.), *Medicina Reproductiva* (2º). México, D.F., México: Editorial Manual Moderno, AMMR.

González-Santos, S. P. (2011). *The sociocultural aspects of assisted reproduction in Mexico* (Doctoral, University of Sussex). Retrieved from http://sro.sussex.ac.uk/7081/.

González Santos, S. P., Stephens, N., & Dimond, R. (2018). Narrating the first "three-parent baby": The initial press reactions from the United Kingdom, the United States, and Mexico. *Science Communication, 40*(4), 419–441. https://doi.org/10.1177/1075547018772312.

Gutierrez Najar, A., Navarro, M. C., Martin, O. E., & Diaz, M. (1992). Fertilización in vitro-GIFT: evaluación de una clínica 1988–1992. *Ginecología y Obstetricia de México, 60*(S1), 100.

Gutmann, M. (2009). Planning men out of family planning: A case study. *Sexualidad, Salud y Sociedad - Revista Latinoamericana, 1*, 104–124.

Hables Gray, C. (2002). *Cyborg citizen*. New York and London: Routledge.

Kably Ambe, A. (1988). Fertilización in Vitro. In J. Delgado Urdapilleta & A. Kably Ambe (Eds.), *V Reunión Anual. Experiencia en México* (Conference Report) (pp. 111–124). México: Instituto Nacional de Perinatología.

Kably Ambe, A., Salazar López Ortiz, C., Serviere Zaragoza, C., Velázquez Cornejo, G., Pérez Peña, E., Santos Haliscack, R., … Gaviño Gaviño, F. (2012). Consenso Nacional Mexicano de Reproducción Asistida. *Ginecología Y Obstetricia De México, 80*(9), 581–624.

Kelly, J. J. (1994). Article 27 and mexican land reform: The legacy of Zapata's dream. *Human Rights Literature Review, 25*(541), 541–570.

Kunhardt Rasch, J. (1999). Presentación (p. 1). In E. Vázquez Benítez (Ed.), *Medicina Reproductiva en México* (Primera). Mexico: JGH Editores S.A. de C.V.

Maldonado, L. D. J. (2017). *Análisis del embarazo adolescente en México 2000–2012 (Tesina).* Centro de Investigación y Docencia Económicas, A.C. Retrieved from http://repositorio-digital.cide.edu/bitstream/handle/11651/2243/158721.pdf?sequence=1&isAllowed=y.

Morton, A. D. (2003). Structural change and neoliberalism in Mexico: "Passive revolution" in the global political economy. *Third World Quarterly, 24*(4), 631–653.

Mulholland, J., Mallidis, C., Agbaje, I., & McClure, N. (2011). Male diabetes mellitus and assisted reproduction treatment outcome. *Reproductive BioMedicine Online, 22*(2), 215–219. https://doi.org/10.1016/j.rbmo.2010.10.005.

Nazar-Beutelspacher, A., Zapata-Martelo, E., & Vázquez-García, V. (2004). Population policies and women's nutrition: A study on six rural communities in Chiapas. *México Agricultura Sociedad y Desarrollo, 1*(2), 147–162.

Pérez Peña, E. (1999). Introducción: La Reproducción Asistida. Cap. 46 (p. 249). In E. Vázquez Benítez (Ed.), *Medicina Reproductiva en México* (Primera). México: JGH Editores S.A. de C.V.

Preston, J., & Dillon, S. (2004). *Opening Mexico: The making of a democracy.* New York: Farr, Straus, and Giroux.

Quark Adams, A. (2019). Outosorcing regulatory decision-making: "International" epistemic communities, transnational firms, and pesticides residue standards in India. *Science Technology & Human Values, 44*(1), 3–28. https://doi.org/10.1177/0162243918779123.

Raju, G. A. R., Prakash, G. J., Krishna, K. M., Madan, K., Narayana, T. S., & Krishna, C. H. R. (2012). Noninsulin-dependent diabetes mellitus: Effects on sperm morphological and functional characteristics, nuclear DNA integrity and outcome of assisted reproductive technique. *Andrologia, 44*(s1), 490–498. https://doi.org/10.1111/j.1439-0272.2011.01213.x.

Rodriguez Argüelles, J. (1999). Prevención de la responsabilidad legal en relación con la reproducción. In E. Vázquez Benítez (Ed.), *Medicina Reproductiva en México* (Primera). Mexico: JGH Editores S.A. de C.V.

Serviere Zaragoza, C. F. (2003). Reproducción Asistida. Indicaciones. Sección XIII Reproducción Asistida. Cap. 39 (p. 285). In E. Vázquez Benítez (Ed.), *Medicina Reproductiva* (2°). México, D.F., México: Editorial Manual Moderno, AMMR.

Tamez González, S., & Valle Arcos, R. I. (2005). Desigualdad social y reforma neoliberal en salud. *Revista Mexicana de Sociología, 67*(2), 321–356.

Tamez, S., Eibenschutz, C., Camacho, I., & Hernandez, E. (n.d.). *Neoliberalismo y política sanitaria en México.* Foro Social Mundial de la Salud. Retrieved from http://medicinaweb.cloudapp.net/observatorio/docs/em/lg/EM2010_Lg_Tamez.pdf.

Téllez Velasco, S. (2003). Presentación. In E. Vázquez Benítez (Ed.), *Medicina Reproductiva* (2°). México, D.F., Mexico: Editorial Manual Moderno, AMMR.

Tlapanco Barba, R. (2003). Contraindicaciones para la reproducción asistida. Sección XIII Reproducción Asistida. Cap. 40 (p. 291). In E. Vázquez Benítez (Ed.), *Medicina Reproductiva* (2°). México, D.F., Mexico: Editorial Manual Moderno, AMMR.

Vázquez-Benítez, E. (Ed.). (1999). *Medicina Reproductiva en México* (Primera). México, D.F., Mexico: Manual Moderno JGH Editores S.A. de C.v.

Vázquez-Benítez, E. (Ed.). (2003). *Medicina Reproductiva* (2°). México, D.F., México: Editorial Manual Moderno, AMMR.

Vázquez-Benítez, E. (2008, Julio–Septiembre). LA Asociación Mexicana de Medicina de la Reproducción 1949–2008. *Revista Mexicana de Medicina de la Reproducción, 1*(1), 3–14.

Welti-Chanes, C. (2014). El Consejo Nacional de Población a 40 años de la institucionalización de una política explícita de población en México. *Papeles de Población, 20*(81), 25–58.

Zander-Fox, D. L., Henshaw, R., Hamilton, H., & Lane, M. (2012). Does obesity really matter? The impact of BMI on embryo quality and pregnancy outcomes after IVF in women aged ≤38 years. *Australian and New Zealand Journal of Obstetrics and Gynaecology, 52*(3), 270–276. https://doi.org/10.1111/j.1479-828X.2012.01453.x.

Zegers-Hochschild, F. (1998). Consenso latinoamericano en aspectos ético-legales relativos a las técnicas de reproducción asistida. Reñaca, Chile, 1995. Red Latinoamericana de Reproducción Asistida. Marzo, 1996. *Cadernos de Saúde Pública, 14*(suppl. 1), S140–S146. https://doi.org/10.1590/S0102-311X1998000500026.

Zegers-Hochschild, F. (2003). *Registro Latinoamericano de Reproducción Asistida 2003*. Chile: Red Latinoamericana de Reproducción Asistida.

Zegers-Hochschild, F. (2011). Barriers to conducting clinical research in reproductive medicine: Latin America. *Fertility and Sterility, 96*(4), 802–804. https://doi.org/10.1016/j.fertnstert.2011.08.043.

Zegers-Hochschild, F., & Prado Aravena, J. (1990). *Registro Latinoamericano de Reproducción Asistida 1990* (pp. 1–13). Retrieved from International Working Group for Registers on Assisted Reproduction IWG website: http://www.redlara.com/PDF_RED/RLA-1990.pdf.

Zegers-Hochschild, F., Galdames, V., & Balmaceda, J. P. (1999). *Registro Latinoamericano de Reproducción Asistida 1999*. Chile: Red Latinoamericana de Reproducción Asistida.

Zegers-Hochschild, F., Schwarze, J. E., Crosby, J. A., Musri, C., & Souza, M. do C. B. de. (2013). Assited reproductive technologies (ART) in Latin America: The Latin American Registry, 2011. *JBRA Assisted Reproduction, 17*(4). https://doi.org/10.5935/1518-0557.20130062.

Zegers-Hochschild, F., Schwarze, J. E., Crosby, J., Musri, C., & Urbina, M. T. (2017). Assisted reproduction techniques in Latin America: The Latin American Registry, 2014. *Reproductive BioMedicine Online, 35*(3), 287–295. https://doi.org/10.1016/j.rbmo.2017.05.021.

6

The Discursive Landscape

Sperm singing out "Help!" in A major; parading couples juggling their children in front of potential patients; the Virgin of Guadalupe helping a mother help her daughter become a mother; a children's book about an "itsy bitsy seed" and how it became a bunny; a bright pink magazine offering its content "Because you want to be a mother"; "Matices", a bank loan that will help you pay for your plastic surgery and your assisted reproduction procedure; a billboard in the middle of the motorway and in the bus stop, both pointing me to the best AR clinic; an article in the newspaper declaring that "Science defeats Infertility"; all these are example of the plethora of elements that make up the Mexican repronational portrait. Some of these messages circulated in the media, were present in public spaces, or narrated at information events and at trade shows. In short, this is the discursive landscape I have encountered during the years I have done fieldwork, and it is what I explore in this chapter. The main question I am asking is: How did the narrative presented in this discursive landscape frame assisted reproduction? How did this narrative contribute to make assisted reproduction sociably usable?

In the previous chapter, I looked at how the universe of assisted reproduction expanded in terms of the number of clinics, practising professionals, eligibility of patients, and existing procedures. Now, I will

© The Author(s) 2020   **193**
S. P. González-Santos, *A Portrait of Assisted Reproduction in Mexico*,
https://doi.org/10.1007/978-3-030-23041-8_6

focus on how assisted reproduction, as a biomedical service directed to the community of potential users and beyond, was presented and promoted in the media and in the non-clinical spaces where service providers and consumers met. The presence of assisted reproduction in the discursive landscape not only contributed to people's awareness of its existence, but simultaneously participated in the process of commodifying assisted reproduction, and creating and expanding a market for it. The narrative constructed in this landscape suggests that this way of forming a family holds an advantage over adoption, since (most of the times) it helps maintain some biological connection with the parents. It frames assisted reproduction as compatible with many established values, particularly Mexican motherhood, to the point that it re-enforces traditional gender roles. In this discursive landscape, the complexities and risks of using ARTs are simplified and downplayed by offering mostly successful narratives and straightforward stories. Finally, while the messages circulating in this discursive landscape cover a considerable range of areas, including success rates, basic elements of the procedures, how to socialise their use, and how to make sense of them, there are very few voices offering a critical perspective, questioning if there could be another way or if there are risks one needs to consider.

The chapter begins by briefly describe what I mean by discursive landscape and offering an overview of the different spaces that make it up. Then, I look at how ARTs are articulated as successful *paranatural* technologies and as a motherly act. The chapter closes with a discussion regarding the commercial touch that has facilitated the commodification of infertility and reproduction and the marketisation of AR services; both processes have shaped the more recent years of the AR system.[1]

---

[1]There are a few scholars whose work dialogues with what is presented here in terms of how people make sense of the different situations and actors that emerge as a result of assisted reproduction, for example Elizabeth Robert's work on Ecuador (2006, 2007, 2016), Melissa Shaw's work on negotiating medicalisation in Colombia (2016), and Mariana Viera Cherro's (2012, 2015) work on how ARTs are viewed and negotiated by clinicians and patients in Argentina.

AR's Discursive Landscape

The Oxford dictionary defines landscape as "all the visible features of an area, often considered in terms of their aesthetic appeal" and also as "the distinctive features of a sphere of activity". Among its common synonyms are: topography, terrain, environment, perspective, and panorama. Discursive is defined as "digressing from subject to subject" or "fluent and expansive style of speech or writing" and within philosophy as "proceeding by argument or reasoning rather than by intuition". Both these definitions help me explain what I mean by discursive landscape: the visible and distinctive narrative features of a particular sphere of activity, where these distinctive features are, many times, aesthetic, fluent, and expansive, and they digress between subjects. Discursive landscapes have a topography, they create an environment, they give a perspective and offer a panorama, and they have textual and visual elements. This concept dialogues with Stephens and Ruivenkamp's notion of "imagescape" (2016), particularly with how they bringing in both visual semiotics and narrative analysis in their visual analysis. For Stephens and Ruivenkamp, images "(1) can circulate as entities in their own right, (2) carry meaning that is independently organised, but (3) convey messages in combination with the frame in which they are presented, and (4) allow for the co-creation of messages through the interpretation of the reader" (Stephens & Ruivenkamp, 2016: 332).

The discursive landscape I analyse in this chapter is assembled of textual, oral, and visual elements, all talking about assisted reproduction through biotechnologies. The visual elements sometimes are two dimensional (printed images) and other times three dimensional (objects present physically). These visual elements can be, for example, the icons attached to the user's profile in online forums, the freebies given out at the trade show, the images used in the power point presentation during information talks, as well as the pregnant women and couples with babies invited to the information sessions as visual proof that ARTs work. These visual messages are both read on their own and in conjunction with the textual and verbal messages that accompany them.

This discursive landscape has several distinctive features. First, it has particular digressions between, for example, the techno-scientific and the moral issues, between the practical aspects and its difficulties, between the risks and advantages of biomedically assisting reproduction. It has certain aesthetic features, not only in terms of the materiality of the discourse (being text, audio, visual, or a mix of all these) but also in terms of their content (being fictional and non-fictional, informative and suggestive, promising and supportive). It has a topography located within the boundaries of the public and semi-public and constructed by the variety of elements that interact, for example, physicians, users, journalists, writers, marketing firms, service providers, patients, mothers, and children-to-be, all interacting in all sorts of relationships like interviewer-interviewee, client-service provider, audience-producer, subject-object of desire, etc. Each of these many voices is presenting its own ideas, generating a particular panorama of what assisted reproduction is and can do, each following a particular set of arguments and reasonings. It is in the discursive landscape where assisted reproduction is confronted with pre-existing ideas regarding reproduction, family, gender roles, nature, science and technology, etc. Like in any other environment, what is happening in one space of the discursive landscape affects the others; at the end, the resulting narratives are constructed through the interaction of the messages presented in each of these spaces. People do not encounter and react only to one of these spaces, but in fact, they interact with many of these daily. What is said and done in this landscape works as a vehicle for the transmission of information, as a way clinics have to promote their services, as a place where people look for and get information, share experiences and opinions, and debate ethical, legal, and social matters. In doing all this, this discursive landscape has forged the usability of ARTs and of biomedically assisting reproduction.

I explored different spaces in this discursive landscape: two support groups, one trade show, several recruiting events, one telenovela, many advertisements, ten years of newspapers, and some television shows.[2] The scope of this discursive landscape reached a diverse audience. For

---

[2]Data was collected from a ten year span (1999/2000 to 2010).

example, by placing adds in the public transport, on prime time television, and in high-rating radio helps disseminate the existence of ARTs and AR services across a wide audience in terms of age, income bracket, and reproductive project. This is important for the process of making them usable, particularly considering a wider understanding of user (Oudshoorn & Pinch, 2003).

Support Groups

> At AMI we feel like a charm, after these sessions someone always gets pregnant. (Field notes, AR user at an information event, 2007)

In 2004, a group of former AR users from Monterey formed a non-profit, non-governmental, national organisation called Asociación Mexicana de Infertilidad (Mexican Infertility Association, AMI). As early adopters, they had lived first-hand the pitfalls of using ARTs with limited information and support. This motivated them to create this support group. They had four purposes in mind: to help people make informed decisions about the procedures used to assist reproduction, as well as help them to consider "other options" for becoming a parent (e.g. adoption), to offer them emotional support through shared experiences, to increase awareness about infertility issues among healthcare professionals as well as among the general public, and to help physicians become aware of the complex and heterogeneous needs people have when undergoing assisted reproduction. In order to address these four objectives, they had a website which housed an online forum, they organised information events, to which they invited physicians, and they organised smaller more informal gatherings. According to AMI's founder, during its most active years, they had chapters in fifteen states and one in the USA, and they had helped over 400,000 couples via their website alone (Interview, 2008). As the opening quote states, being part of AMI helped.

As stated in their adverts, information sessions were organised to help people "achieve pregnancy when it seemed impossible". These were space where physicians and people seeking medical help could speak about infertility and the ARTs in a context other than the consultation setting and the patient-physician dynamic. The event included a presentation by an AMI member followed by a presentation by the physician. After the coffee break, the physician would sit with the audience to listen to their questions and experiences in a more horizontal collective interaction. The people who went to these events were usually heterosexual couples. Although all were there for the same purpose, couples kept to themselves, holding only minimal interaction between couples.

The interaction at the information sessions contrasted with the interaction at AMI's website. AMI's webpage offered a range of sections with information about infertility, assisted reproduction, and the association, as well as a very active online forum. The online forum was frequently used, but mostly by women,[3] even in the section specifically labelled for men there were more women than men. Forum members accompanied each other's progress, shared their experiences and feelings, commented on their diagnosis and treatments, exchanged information about doctors, clinics, prices, symptoms, procedures, latest technology, and gossiped about celebrities using assisted reproduction. As I explored this site, I found that it was common for people living in different cities and different countries, to compare how things were done in their different local clinics. The heterogeneous information exchanged through AMI's forum helped users and potential users gain various types of knowledge concerning infertility and assisted reproduction. It allowed them to share and discuss different ideas and experiences and find ways of making these procedures acceptable and usable. According to what they said in the forum, all participants were either currently undergoing some sort of treatment to assist reproduction, had recently undergone treatment, or were about to begin.

In the forum, users interacted with fellow users and not with physicians, like at the information session. The forum had one section for those undergoing treatment and one for those already pregnant; under the former, there were three more options, one for those using ARTs, those using ARTs with donation, or those who were adopting.

---

[3] By this, I mean that most were self-identified as women both in their textual interaction and in the name they used.

People would participate in the section that corresponded to the stage they were at. However, within the assisted reproduction section, there could be a person undergoing artificial insemination talking to someone undergoing in vitro fertilisation (IVF) or two people using IVF but one with PGD and the other about to freeze their embryos. The variety of specific procedures being commented upon and the diachronic structure of forum made it possible for people to be constantly reminded of what they had just gone through, what they could await, and what they could never accept.

Another support group was Fertired, established online in 2007 by two women who had been involved in the assisted reproduction since the eighties, when the first clinics were being established, one as a communication's manager and the other as a psychologist. The objective of Fertired was to create a "fertility community" which included all those involved in infertility: patients, psychologists, and biomedical specialists, men and women, singles and couples. Its name, which is a word built from the conjunction of the word fertility (ferti) and network (red), their slogan, "Encuentros fértiles sobre infertilidad" (fertile encounters about infertility), and their logo (an abstraction of three people intertwined with no gender indication, so it could be a group of women, a group of men, a group of men and women, or even a family), all stressed the idea that, on the one hand, it takes collective work to achieve pregnancy with assisted reproduction, and on the other that empathic support helps deal with infertility. Like AMI, they too believed that, if the biomedical community understood the sociocultural and emotional aspects of infertility and assisted reproduction, they would provide a better service. Therefore, they offered workshops to develop sensitivity among the clinic's staff and group counselling for the biomedical team. They also offered information sessions as well as emotional support groups for their patients.

At both these forums, participants debated about meanings, struggled to re-signify kinship, strived to convince themselves that their decisions were correct, looked for approval, gave support, and sought advice. For example, in the following quote a woman states how she resolved her doubts about using gamete donation for fear of having no resemblance between her child and her husband and herself:

I think that thanks to forums like this one and the information it spreads out, these thoughts [fear of lack of resemblance] start to fade away from the minds of those who are in search of having a baby, and this way the advances in science are accepted with greater ease. (Forum, AMI, 2009)

As that quote exemplifies, these interactions offered people tools to deal with specific situations. Another example was a commonly circulated phrase that helped people using donated gametes, embryos, or adoption deal with issues of resemblance: "It will not have your eyes, but it will have your gaze; It will not have your mouth, but it will have your smile". This mantra-like phrase was constantly repeated by people who sacrificed the genetic and maybe even gestational tie by strengthening the nurture tie and assigning to it elements of resemblance (Becker, Butler, & Nachtigall, 2005). This phrase was used in the forums, at the information sessions, and even during interviews, I had with patients.

In these interactions, there was also a process of establishing what would be accepted and what would not. For example, on one occasion, a woman posted a long message describing whom she and her husband were and how they would like to help those who could not conceive to become parents. She offered the forum members her womb in exchange for money. The president of AMI promptly responded leaving the post as a way of showing "to our AMI friends what should not be done". What this woman was offering was illegal in Mexico (paid surrogacy). With this case, as with others concerning adoption and gamete donation, AMI lent its support for certain types of actions (i.e. legal procedures) and condemned others.

Another issue that was discussed, negotiated, experimented, and socialised in these spaces was who to tell and what to tell. People using AR services constantly need to figure out how to share their experience and their story and with whom. Frequently, they stated that discussing these concerns among the AMI community helped those who had not yet decided, to make a decision regarding how much to share and with whom. In summary, through the social interactions in forums like these, people acquired and exchanged information, established what was acceptable and what was not, negotiated and reassigned meanings, and shared feelings and ethical concerns, and by doing so, they participated in the process of making certain uses of assisted reproduction a socially accepted way of forming a family.

Today (2019) both Fertired and AMI's blog and information events have ceased to exist. Their founders and managers have moved onto other projects. The people behind AMI, for example, have created Proyecto BB en casa (Project BB at home), a programme designed to help people finance their treatment which they carry out in conjunction with certain AR clinics and with the pharmaceutical company Ferring.[4] One of the founders of Fertired set up a specialised pharmacy called Fertifarma. Until very recent, finding the different drugs used in the different AR protocols could be very difficult, Fertifarma took advantage of this situation and transformed it into a profitable business.

The Trade Show

The year was 2007. I was standing in the middle of one of the large exhibit halls of Mexico City's World Trade Centre carrying a heavy pink bag full of information: USBs in the shape of sperm, a pen that when tilted had sperm floating to the ova, post-its and colourful pens, folders with explanations about the procedures, about financing options, and dozens of cards with the contact details of clinics and doctors. I saw dozens of couples holding hands and grasping the pamphlets and booklets they had collected at the different stalls. There were light boxes with blown-up images of glossy colourful eight-cell embryos, of syringes puncturing ova, of swimming sperms, and of happy families. I spoke to doctors in white coats and heard them talking to couples. I watched how men and women dressed in suits were handing out stuff. I also saw a masseur's table waiting to be used, a stall selling non-stick chemical-free pots and pans, a miscellaneous collection of objects, images, people, and services that were being related, just by being there,

---

[4]Ferring is a pharmaceutical company that began operations in Mexico in 1996 and since has had great local grown. http://esperandote.net/fertilidad/bb-en-casa/.

to infertility and assisted reproduction. Standing in the middle of the venue I could smell the aromatherapy, I could hear a pandemonium of overlapping explanations of procedures, diagnosis, and financing schemes uttered from all around. Far off to the left, one clinic's video was a clip of animated sperms singing an adaptation of The Beatles' hit song Help!: the sperm was asked the doctor to help them fertilise the ova. In the centre, another promotional video with a female narrator talking about the clinic's facilities and the procedures they offer. The combination of all this offers an eclectic, sonorous, smelly portrait of what infertility and assisted reproduction entail. It is a portrait with hues and tones related to biomedicine and non-biomedical health methods, with elements making reference to economy, to high technology, and to strong emotions. All this was Expo Fertilidad.

Expo Fertilidad was a trade show dedicated to fertility,[5] created and directed by a woman in her late thirties with a carrier in public relations and marketing.[6] It took place a total of five times (2007, 2008, 2010, 2011, and 2012), all at the World Trade Centre in Mexico City. Each year the event was announced several months in advance, with bright pink posters placed on bus stops, billboards, valet parking stands,

---

[5]Events similar to this one have taken place in other countries, for example the National Infertility Day in the UK, the Fertility Expo, and the New Beginnings fertility conference in the USA. Some of these events are for free and organised in conjunction with support groups while others are done with the support of clinics. However, the organiser of Expo Fertilidad claims this as the first event of the sort to be held in Mexico and in the rest of Latin America. If other health-related patient led conferences take place in Mexico, they never receive the same amount of media coverage nor are announced in so many public spaces as this one.

[6]During an interview, she told me the story of how she came up with the idea of Expo Fertilidad: she had been working at a parent's magazine (called Padres e Hijos), when she was asked to organise a sweepstake, as a way to promote an infertility clinic. The sweepstake was called "You haven't been able to be a mother? We make your dream come true" (Padres e Hijos, 2005 XXVI [12]: 58). The idea was that people would send in letters telling their infertility story accompanied by their medical diagnosis. The case with the best possibility of achieving success would be offered a free AR cycle at the participating clinic. "We received about 1000 letters. For other contests, for example, having your baby's face on the front cover of the magazine, I would receive about 3000; so you can see the proportion [...] my readers were supposed to be parents already; probably some were just aspiring to be parents and others knew people who were trying to become parents". The unexpected number of letters received as well as their content made her acknowledges that infertility and assisted reproduction were becoming important topics for which people had little information. The high response and the clear need for information gave her the idea of a trade show where patients/consumers could meet clinics/service providers.

magazines, and digitally. Days leading up to the event, Expo Fertilidad was everywhere, calling out the event's slogan: "Queremos que alcances tu sueño" (We want you to reach your dream). For the first year, the media was invited to cover and participate in the inauguration of the event.

The 2000 pesos ticket covered the entrance to both the exhibit hall, where the twenty-four stalls were placed, and the ten conferences and four workshops. The cost drastically dropped the following years; by the last edition, the price was approximately 150 pesos per person. In 2007, ten stalls belonged to fertility clinics (only one was of a clinic located outside Mexico City); the remaining fourteen were promoting non-biomedical methods to improve fertility, from aromatherapy, Chinese medicine, acupuncture, and the use of magnets and massages, to food supplements and cooking utensils for a healthy diet, including machines to measure hormone levels, temperature and blood pressure, and an array of services and things unrelated to infertility and assisted repro-duction. Noticeably, no adoption agency was present and the only phar-maceutical company was Merck Serono, which had its name on banners in the conference hall and in most of Expo Fertilidad's materials (bags, name tags, and furniture), but no stall. Although in the exhibit hall all these ways of dealing with infertility were standing side by side, they did not come across as equal. While the stalls put up by the AR clinics were massive (some two stories high) and with a lot of images and videos, the non-AR ones were smaller and less flashy. Clinics arranged their stalls either like a mini-clinic (with a reception desk, a waiting area, a consul-tation area, and a replica of a laboratory), or like small conference spaces in which they gave talks that could be heard all around the venue. At the stalls were physicians in white coats or in suits, young women dressed in uniform, and sometimes the nurses of the clinic, although they usually did not present themselves as nurses. The role of the young women was to attract people's attention, give out memorised informa-tion and souvenirs, write down people's contact information, promote discounts, and hand out brochures and freebies. The talks and the work-shops covered various topics related to assisted reproduction such as the causes for infertility, the emotional effects of facing it, the different pro-cedures used to overcome it, and the elements that constitute a good fertility clinic. In addition to the biomedical physicians, there were also few psychologists, andrologists, biologists, and non-biomedical healers invited to speak at the different events, although more time was allocated to the gynaecologists. Attendees, who were mostly couples, were able to compare prices, procedures, and styles between clinics and doctors. Although this comparison was also done in the forums, here it was taking place in a market-like setting, where discounts and promotions were put forth to convince people that theirs was the best option.

Like AMI and Fertired's forum, Expo Fertilidad also stopped. Every year there were fewer and fewer attendees. The organiser attributed the decline in attendance to the fact that people had more information: "I think that doctors are more aware that people need information and are doing something about it, either organising information sessions, or advertising more, or something, but I feel they are a bit more informed than last year". Interestingly, she never mentioned her own efforts (i.e. the previous events) as being responsible for her hypothesis that people had more information. The only events that persist today are the recruiting events that I will describe in the following section. The last I heard from the organiser she was planning on moving onto fertility preservation, organising events to convince university students to freeze their eggs.

Recruiting Events

Some clinics held weekend morning events at hotel conference rooms. They advertised these events as free information sessions where you could also get discounts for future consultations and treatments. Although the different names clinics use to refer to these events (i.e. seminar, informative chat, informative session) suggested, they were only to inform people, and according to the organisers, these information sessions were not for profit—"they are completely for free, we actually have to invest in them"—the structure of the event, its visual arrangement, the images presented, the speakers' attire, and the content of the talks all followed a well-planned, orchestrated, and performed marketing scheme to recruit patients. Through these events, clinics created a database of potential patients which they would contact days after the event to remind them that they were eligible for a discount in their first consultation and they would offer to book their first visit.

These events usually followed a similar structure. The physical arrangement of the room resembled the setting of any other formal professional gathering, a podium framed with banners on both sides, chairs facing the podium, coffee and cake in the back, and a desk where you register your contact information. Attendees were usually couples. They would not interact among them, almost the contrary. It looked like the couple would stand or sit close together, hugging or holding hands forming a unite. Organisers usually dressed in dark grey suits (sometimes even a uniform) with no evident features in their attire to indicate if they were physicians, part of the PR, or economists (nobody wore white coats). The first to speak was a male physician (the medical director), and he would succinctly introduce the general biomedical aspects of infertility and the different ARTs and gave a general overview of the clinic. Sometimes a psychologist would moderate the session and talk a little bit about the emotional aspects of infertility and the use of assisted reproduction. Then, a female speaker presented the financing scheme they offered to cover the cost of the procedures, followed by the Q&A session. Attendees were invited to ask the doctors all their questions. People would ask doctors for their point of view regarding their case, and they would share their medical history and test results, frequently using some type of technical jargon and information (e.g. the biomedical name of their condition or procedure, the names of medicines, their hormone levels). The event usually closed with the display of success: presented as a "surprise", pregnant women or couples with their babies in their arms would walk into the room, pushing prams, juggling one or two babies, blanket, and toy. They would take the podium, show their dressed up baby, and tell their moving story. This was the only moment in which the voice of the AR users was heard in these events. The user represented success. Evidently, there was never a non-success story told.

The key point of these interactions was that they were not in the consultation setting. In other words, they were not a patient sitting in front of a physician in the clinical encounter, nor two patients face to face in the waiting room of the AR clinic. This difference is important because there are actors that appear in these events that could be confused with actors in other settings. Let me paint a hypothetical yet common scenario as a simplified example. Although in both settings, in the clinical and in the non-clinical, Dr. X and Mrs. Y meet, following Annemarie Mol's ontological perspective (2002), Dr. X and Mrs. Y are not the same actors in each of these cases. They are different because they are the result of different relations. At the clinic, Dr. X is dressed in a white coat within an aura of expertise, and he is probing Mrs. Y's body with tools that have encrypted knowledge, asking her intimate and many times uncomfortable and painful questions, offering her few answers, and receiving a payment. All this places Dr. X in a particular powerful position over Mrs. Y, who has less medical knowledge, limited economic resources, and has a profound pain and desire for a child. This clinical scenario is very different from the non-clinical one offered by these vents. There, Dr. X is wearing a suit, instead of the white coat, standing at a stall waiting for Mrs. Y to arrive. When she does, after visiting other stalls, Dr. X has to convince her (who is now an attendee not a patient) that their clinic is the best that they are the best option for the Y couple to take their business of wanting a child. The traditional "patient-physician relationship" does not happen in this setting. People are not patients, they are attendees empowered because they are a consumer. Physicians are experts, yes, but selling a service and competing with others who equally claim to be the best. These interactions offered new ways of explaining things, of negotiating between competing values or actions, and they were spaces for experimentation on how to make sense of assisted reproduction and how to explain it to others. This was all a new for both Dr. X and Mrs. Y.

Some of the doctors I talked to at the trade show commented on this new type of interaction. One said that at first, he felt odd being in a situation in which he had to "sell himself that way", for he was used to "seeing patients in his office and not standing in a booth"; however, by the end of the third day, when the event was over, he said it had been a good experience. Another doctor agreed and added, with an exhalation of exhaustion, that he had been talking to patients non-stop for two days. Both said it had been a good experience and participated in the event in the following two years. Other clinic directors did not like the market-like character of the event, and, although some participated as speakers, they never attended with a booth or only did so the first year.

Children's Books

Every year there are more children born after using ARTs. As these children grow, parents have to face the decision concerning what to tell them regarding their origin story. This has given place to a new site where assisted reproduction is being disseminated and made usable: children's books. In 2005, one of the AMI founding members, Carmen Marinez Jover published both in English and in Spanish, a book on egg donation called "An itsy bitsy gift of life: an egg donor story" (Un regalo de vida chiquitito: un cuento de donación de óvulos). This book tells the story of two rabbits, Paly and Comet, who want to have bunnies, but after a year, the doctor tells them that Paly no longer has "the itsy bitsy seeds" she needs to make a bunny. This made her very sad. One day, an unknown lady rabbit knocks on their door and offers to give her one of her "itsy bitsy seeds" since she had many and Paly did not. The image shows a lady rabbit at the door with bunnies crawling all over her. Paly was very happy with the gift this unknown lady rabbit gave her. In the image, we see her holding a test tube with a red seed at the bottom and ribbon on the top and Comet stretching his hand out with a blue seed. In the next page, Comet, the male rabbit, explains that they need to put both "itsy bitsy seeds" in the tube to create the bunny. With no further explanation, in the next page we see Paly with a swollen belly, knitting, while Comet brings her nutritious food. The next page shows the newborn bunny laying in her crib. The story ends with the bunny girl as a toddler and all as a happy family. This story omits the medical aspects such as the doctor and the clinic. It would be useful to see how other stories present these processes and how parents are telling the origin stories of their children to them.

Media and Advertisements

By the turn of the century, assisted reproduction was already on the streets, as billboards in the middle of motorways and in backstreets, as advertisements in public transport such as the Metrobus,[7] and as commercials and infomercials in magazines, newspapers, on television, and radio. As the years went by, between 2000 and 2010, the number of newspaper articles talking about assisted reproduction in the three major newspapers of the country (*El Universal, El Reforma,* and *La Jornada*) increased.[8] Just as an example: in 1999 *El Reforma* published six articles on infertility and assisted reproduction, one year later, in 2000, this number increased to twenty-four. This increase also happened in the other newspapers, as well as in magazines, on television, and in radio. Women's magazines published either single articles or entire issues dedicated to infertility and assisted reproduction[9]; daytime talk shows invited AR specialists and users to discuss these services, science communication programmes explained the functioning of ARTs,[10] and radio shows presented everything from the science behind it, to the

---

[7]The Metrobus is a confined bus system that has served Mexico City since mid-2006. In 2017, estimates said that with its six lines, it moved an average of 1.3 million users daily.

[8]In conjunction, these newspapers cover the political spectrum as much as possible: one has the largest circulation in the country and actively attempts to follow a non-partisan editorial line (*El Universal*), another represents a right of centre perspective (*El Reforma*), and the third one a left of centre (*La Jornada*). I used their online databases and search engine using the terms infertility, fertility, sterility, assisted reproduction, and surrogacy and chose the pieces that dealt with human reproduction. In all these cases, I specifically focused on the process of naming, socialising, and establishing of infertility and assisted reproduction. Six themes were found to be recurrent in most discourses, so they were selected as elements to guide the overall analysis: (1) definition, incidence, and causes of infertility; (2) definition, success rates, and side effects of the procedures; (3) criteria for clinic, doctor, and patient eligibility; (4) sources of information; (5) the interaction between gender roles, infertility, and assisted reproduction; and (6) ethical issues. In addition to these themes, I also analysed the discourse strategy employed (use of reference to experts, testimonies, statistics, numbers) and the tone in which articles were written (whether they were presenting great concern, concern, were neutral, or were promising). It is important to clarify that the vast majority of the newspaper and magazine articles are written by journalists who were not specialised in the field of health, medicine, or science communication; hence, it is not uncommon to find errors, for example in the names of procedures (e.g. instead of ICSI, they say IXI).

[9]*Salud y Bienestar,* 2005; *Mari-Claire,* 2006; *BbMundo,* 2006; *Deep,* 2006; *Visión Universitaria,* 2006; *ABC,* 2006; *Nexos,* 2006; *Fernanda,* 2006.

[10]*Diálogos en Confianza,* channel 11, 2003; *n Vitro,* channel 11, 2007.

legal issues that emerge from its use.[11] Much of the information given out in these different advertising spaces seemed to be developed by marketing companies. They usually presented the same sort of uncritical narrative, framing ARTs as highly successful and straight forward procedures.

In the USA, where the idea that patients should be treated as consumers who know what is best for them, the AMA and the ASRM have established guidelines regarding the advertisement practices for AR clinics. Nonetheless, studies have shown that these guidelines are not always followed. In 2005, Huang et al. found that "[i]rrespective of the practice setting or university affiliation, the overall quality of the fertility clinic websites is poor, failing to meet most of the AMA Internet health information guidelines" (p. 114) (see also Abusief, Hornstein, & Jain, 2007). Mexico's law concerning health-related advertisements (Reglamento de la Ley General de Salud en Materia de Publicidad), although updated in 2014, is very poor in the area of advertising healthcare services and even more so when it comes to advertisements in social media like Facebook, and the Mexican Association of Reproductive Medicine (AMMR) does not have any advertising guidelines (Dyer, 1997). Therefore, it is not surprising to see that all clinics employ a persuasive narrative in their advertising campaigns (Barbey, 2017; Hawkins, 2013, 2017; Madeira, 2013)

The most common media coverage of assisted reproduction was through articles which were a mixture of science communication and publicity, or maybe I should say publicity disguised as science communication. Articles claimed that infertility is at a rise. The cause for this rise was described as couples postponing pregnancy more than before, exposing themselves to more infectious diseases and to a prolonged use of contraceptives. Articles would also mention other causes for infertility such as alcohol abuse, smoking, and habits that affect sperm production (e.g. tight clothes, saunas, and exercise). They presented a message of certainty-of-hope. Articles urged readers not to worry since ARTs were successful and could help couples, regardless of their condition, to have the

---

[11] *Los Abogados*, MVS Radio, 2008.

desired children. They gave ample voice to the "experts". Articles usually presented the views of more than one AR specialist, including the name of the clinic where they work, and frequently stating their success rates.[12] Many also presented an economic talk: reference to prices, the different banks that offered credits for these procedures, and the different payment packages the clinics offered (e.g. two cycles for the prices of one, or pay two and the third only if it is successful, etc.). Articles rarely mentioned the problems that can emerge from ARTs, for example, that many people found procedures painful, that engaging in assisted reproduction is time and resource consuming, that the pilgrimage between clinics and treatments is stressful and difficult, and that many times these procedures were not successful. The few times when articles presented an ethical or moral concern was when they were also talking about cloning, genetic manipulation, or abortion. In these cases, the tone of concern was usually directed to these three issues more than to assisted reproduction. The mixture of all these elements creates a narrative where ARTs are presented as a biotechnological, scientific, acceptable procedure that will be successful in all the cases in which it is used.[13]

For a couple of years (2008–2010), the creator of Expo Fertilidad also published a specialised magazine called "Tu Fertilidad: porque quieres ser mamá" (Your Fertility: because you want to be a mother). This small format magazine was clearly targeted to women, as can be inferred by the images and texts in the covers. Out of 15 issues, five had a white happy looking couple on the cover (one was with a pregnant woman), seven had single young smiling women (three were pregnant), one had a family, one had a baby, and one a baby and its mother. Not one had a single man, a man with a baby, a same-sex couple, or non-white person. It was also clear from the text in cover, with statements directed to women's

---

[12]It is important to underline that these articles were not always only about private clinics. There were also articles talking about the AR programmes offered at public clinics like INPer and ISSSTE.

[13]This is similar to what Aditya Bharadwaj talks about (2000) when he analysed the "narrativisation of infertility" in the Indian mass media. In his paper, he showed how the media combines "scientific and journalistic styles to create an institutional advertisement which makes the subject—conquering infertility—appear magnificent and yet attainable" (p. 70).

concerns such as: "A helping hand for the stork", "Be positive, you can be a mother", "Obsessed with having a baby? Find out how to solve it", and "Everything you need to know about ovo-donation". Nevertheless, inside the magazine it did include articles on non-heteronormative couples using assisted reproduction ("Sexual diversity: AR for everyone", 2010).

Although the focus of the magazine was infertility and how to overcome it, the perspectives talked about in the different articles were varied. There were articles on biomedical procedures such as Intra-cytoplasmic sperm injection (ICSI) and assisted hatching and on socio-biomedical options like ovo-donation (2008, 2009) and surrogacy (2010). The magazine also included articles on treatments that were not commonly offered at AR clinics ("From the naturist to the clinical: pro-fertility treatments", "Find out about alternative therapies", 2009), such as the use of psychotherapy and aromatherapy ("Psicoaroma: helps you unblock", 2010), the importance of nutrition and exercise ("Food that improves reproductive capability", 2008), the power of flowers, and even humane birthing. In 2010, the last year the magazine was published, it touched upon three issues that were not commonly addressed: the option of not having children at all ("Living without children: the DINK, you chose", 2010), the use of egg cryopreservation ("Vitrify your ova and ensure your maternity", 2010), and adoption ("Adoption: another way of forming a family", 2010).

Specialised pharmacies (Fertifarma), specialised products, and AR clinics advertised in this magazine both with paid ads and by participating in the directory. Physicians, however, advertised by giving interviews that were presented as articles ("Interviews with the best specialists", 2008; "Meet three senior experts and three upcoming ones", 2008). This is another example of how, in some cases, the clinic and the physician have separated. As detailed in the previous chapter, many AR clinics now have names, logos, websites, and a staff of physicians that together follow the patient's protocol. It has happened that a clinic goes on caring for patients while the physician who had originally established it has moved on to another clinic. This is a new way of practising medicine in México. It contrasts with how private medicine in all other specialties (maybe with the exception of some plastic surgery practices) is usually carried out: patients seek the attention of specific doctors, they go to their office which do not have a name other than the doctor's name, they seldom have logos, or such a large staff.

Overall, the message in these publications (newspaper, magazines, and the specialised magazine) offered a very positive and uncritical perspective of assisted reproduction. This meant two things. One, advertisements tended to present extremely high success rates (several stating between 90 and 100% success rates) and quoting prices that always fell short of what the patient paid at the end (among other reasons, because it is not possible to predict the amount of hormones needed for the ovarian stimulation, and it is these medicines the ones that increase the price of the procedures). Two, by presenting these procedures as successful, resulting in happy endings, it also meant that they were presenting assisted reproduction as a good and valid option.

In addition to the press and public-space advertisements, I am interested in how assisted reproduction was depicted in television drama series. Drawing on the work of social science scholars who have analysed how infertility and ARTs have been depicted in the press (Bharadwaj, 2000; Michelle, 2007; Mulkay, 1994a, b; Shalev & Lemish, 2012), in fictional productions, like movies (Maher, 2014; Nerlich, Johnson, & Clarke, 2003), and in reality TV (Osborne-Thompson, 2014; Weihe Edge, 2014),[14] in the following section I look at how assisted reproduction was presented in a high-rating telenovela-style series.

---

[14]Shirley Shalev and Dafna Lemish (2012) explored the way Israeli's Hebrew press depicted (in) fertility and ARTs between 1995 and 2003. Their analysis concluded that the press framed childbearing as the most valued and sacred contribution women could make to the national collective. It presented motherhood as a glorified public role, prioritised above and beyond personal ambition, needs, and other life goals and as an essential condition for social and familial acceptance. It established a clear preference for the genetic link between parents and their children. And it depicted ARTs both as a source of Israeli pride, due to its techno-scientific component, and as an accepted method to become a mother. The sum of these narratives resulted in a particular way of framing women who use ARTs when facing infertility. They were commonly framed as altruistic individuals willing to accept any physical, emotional, or economic cost in order to bring a child into the world. Although the Mexican press also favoured the genetic tie, also framed motherhood as women's highest and most noble purpose in life, and also presented ARTs as an acceptable means to achieve biological motherhood, the countries' particular political and social contexts result in different policies and reasons behind why these framings are thus. Israel's history (shaped by, among other things, the Holocaust, the ongoing conflict in the Middle East, the demographic threat, and the worldview of Judaism) has brewed an important cultural pressure to produce Jewish children in order to guarantee the survival of the Jewish people. This has resulted in a fertility policy which actively promotes reproduction offering women free and unlimited access to fertility treatments in order to achieve the normative family size of four: a couple with at least two

Telenovelas and Other Dramas

Telenovelas are fictional stories that aim at representing everyday life following a melodramatic structure and with a primarily commercial purpose. Viewed by millions of people daily, both in Mexico and across the globe, they have the power to make, remake, and circulate meaning. They are narrative spaces for expressing, recognising, and recreating culture, where individual and collective imaginaries are built and transmitted, where beliefs and expectations are validated, class structure is stereotypically represented and reproduced, gender roles established, portrayed, and perpetuated, and cultural and moral values promoted (like virginity, machismo, marianismo, corruption, etc.) (Orozco Gómez, 2006). In sum, they play an important role in the construction, production, and perpetuation of culture, and they are a central part in Mexico's tradition of "governance through culture" (Beard, 2003; Flores Palacios & Sánchez Santana, 2006; Lewkowicz, 2015; Orozco Gómez, 2006).

---

children. Mexico, on the other hand, still carries the population bomb discourse on its shoulders; thus, it does not promote childbearing in this way. Although both the public and the work-related health systems offer assisted reproduction, they justify the expense with two combined reasonings: these services are part of the training they offer med students, and they are caring for people's needs.

Maher (2014) offers an interesting analysis of how, in some movies, the heteronormative and genetic-centred reproduction script is upheld through assisted reproduction. She looks at three romantic comedies with female leading roles: Baby Mama (2008), The Switch (2008), and The Back-up Plan (2010). The main argument she found in these films praised women's economic and reproductive independence while at the same time re-enforcing the idea that women, all women, eventually feel the tug of the clichéd "biological clock". Situated within the consumer culture of the USA, where "choice" is presented as a given, all three leading characters resort to ARTs to achieve newly awoken desire to become a mother, all three achieve pregnancy in their first attempt, and all three end up married with the man they love, who also happens to be the biological father of their child.

Brooke Weihe Edge (2014) and Heather Osborne-Thompson (2014) analysed the unintentional use of serial reality television as a way of introducing infertility and IVF into the public conversation. They highlight how these TV shows articulates melodrama and "can-do-ism", suggesting that the solution to the problem of infertility is the use of all available technologies for as long as it takes.

The issue of reproduction has a long history of being presented in the media (for a concise overview, see Basten, 2010). In Mexico, as part of the family planning campaigns of the seventies and eighties, a series of pro-social telenovelas were produced by Miguel Sabido for Televisa. These were a key part of in the larger coordinated effort that brought together the media, public health agencies (the Health Ministry and the Education Ministry), and voluntary groups with the purpose of slowing down population growth (see Chapter 3, Managing Reproduction). These telenovelas promoted the family planning agenda which included reducing the number of children families should have, strengthening family-oriented values, and responsible parenting.[15] The educational success of these telenovelas was measured in terms of the number of people actively interested in family planning by keeping record of those soliciting information, the increase in the sales of contraceptives, and the number of women enrolled in family planning clinics (Basten, 2010).

Most telenovelas, La Rosa de Guadalupe (Guadalupe's Rose, La Rosa) being one of them, are about the life story of "common" people; therefore, looking at how they depict life with infertility and the use of ARTs offers a different sort of narrative than the one offered by celebrity news or scientific reports. The general plot line is that a common everyday person faces a devastating problem (from disease, to alcoholism, drug-dealing, infidelity, bullying, infertility, etc.). In order to solve the problem, the person tries everything, including going to pray to the Virgin Guadalupe to her shrine. As a way of showing the viewer that the prayers have been heard and the problem will be solved, a white pristine rose inadvertently appears on scene, this indicates to the viewer that the prayers have been heard and the Virgin will help. The problem is resolved, and a voice-over offers a socio-moral reflection on the problem addressed.

---

[15]The telenovelas produced by Miguel Sabido were Ven Conmigo (1976), Acompáñame (1977), Vamos Juntos (1979), Caminemos (1980). See Chapter 3: *Managing Reproduction* for more on this.

La Rosa (written by Carlos Mercado for Televisa) is an hour-long unit-based melodrama or telenovela series, aired nationally and internationally since 2008 by Televisa—"the largest telenovela producer in the world, exporting beyond geo-linguistic borders throughout Europe and Asia" (Lewkowicz, 2015: 275).[16] La Rosa falls into a particular type of telenovela, one that in addition to entertaining has an overt purpose of participating in the social discussion regarding certain important social problems. The episode I address here was called *Una Luz de Esperanza* (A Light of Hope), it was aired in 2008, and it tells the story of Roberta and Santiago, a young middle-class couple recently married and facing infertility.[17]

It should be noted that the Virgin of Guadalupe is Mexico's principal religious figure. She has been central to Mexico's history and identity and has been used as a tool for social cohesion for over five hundred years (Finkler, 1994; Napolitano, 2009; Pastor, 2010). During the colonisation process, for example, she was used to convert locals to Christianity; later, she was the bastion of the Insurgent's army giving birth to an independent nation, and she has been central in the integration of the monogamous patriarchic family (Pastor, 2010). The shrine dedicated to her, located in Mexico City, is one of the most visited Catholic sites in the world, receiving over seven million just on 12 December, which is the day she is celebrated (Ahedo, 2017). But the Virgin represents more than just a religious figure, she also embodies the ideal stereotype of how Mexican women should be, what is called marianismo[18] (Pastor, 2010). As La Rosa de Guadalupe's closing song states, the Virgin is "the mother of our country, the mother of us all" (Finkler, 1994; Napolitano, 2009).

---

[16]By this time, Televisa was in a franc competition with TV Azteca for telenovela audience. However, whereas Televisa claimed to be the dream factory "la fábrica de sueños", Azteca—together with Argos Productions—aspired to a more intellectual audience saying they produced telenovelas that "made you think" (Lewkowicz, 2015).

[17]Soap operas and drama series (telenovelas) included assisted reproduction in their plot line.

[18]The Virgin of Guadalupe is the Mexican version of the Virgin Mary; hence, "marianismo" is the theoretical concept that encompasses the ideal stereotype of the Mexican woman, and it is derived from the Virgin Guadalupe-Mary.

ARTs: Successful *Paranatural* Technologies

La Rosa de Guadalupe. Scene 37. An image of the Virgin of Guadalupe and the rose. Voice over: "We women are privileged because we have the gift of being mothers. There are occasions where this is not easy to achieve, but medical science has found many roads to help women become mothers. Love and medicine give us women the light of lost hope"

At the recruiting events clinics present the procedures as a high-tech successful procedures that would help them (as they say): 'create a life through love and science'. (Field Notes, 2010)

as human beings we are not very good when it comes to procreation, every try holds only a 23% chance, it is never 100%, because it is a selective process, and this is something established by nature. (Dr. Information Session, 2008)

Everything that nature can no longer do, we do. We make women ovulate by giving them the same hormones their bodies naturally produce, but these are artificial [...] they are harmless and I promise you, the injections do not hurt [...] We cut the line of genetic diseases being passed down [...] we do all that nature cannot [...] we chose the best embryo. (Dr. Recruiting Event, 2008)

In reality, assisted reproduction implies trying to imitate nature. (Dr. Gutierrez Najar at the Chamber of Deputies, 2001)

They substitute the biological processes that originally take place in the organism (like the maturation of sex cells and fecundation) and that now can take place outside the body in lab conditions that faithfully reproduce the organic environment. (Newspaper article, Flores, 2007)

...with the exception of the cases in which the woman has a small uterus, there is no reason why all the people here could not go home with a baby, everything has a solution. Almost all couples can achieve pregnancy [...] 96%, or even more of the couples that attend INGENES, solve their infertility problem [...] there are only two things needed for success: 1) a good embryo, 2) a good endometrium. (Dr. Recruiting Event, 2008)

96% of the cases that enter the program achieve pregnancy. Not all in the first try, it is not magic, but if they persist and follow the programme, I assure you that you will have your baby. (Dr. Information Event, 2007)

Science defeats Infertility. (Newspaper article, Cerón, 2007)

These quotes an excerpts, placed side by side, offer a clear example of the discursive landscape I am referring to: a conjunction of visual and textual remarks regarding assisted reproduction and infertility, stated by physicians, reporters, and publicists, in a variety of public and semi-public places, with promotional purposes, following a market rhetoric, and with very little critique or counter-narratives. This discursive rhetoric describes ARTs as high-tech, successful, paranatural technologies that go parallel to and beyond nature. Doctors and journalists describe ARTs as a group of "harmless" procedures that sometimes "imitate" nature by "faithfully reproducing" the organic environment; other times "substitute" some biological processes; and all the time they "improve" the outcome because they "defeat" infertility. Because ARTs are "imitating" the natural process, one must not worry about negative consequences; at the same time, ARTs are capable of going beyond nature, with them we are able to select the best and eliminate the undesirable; thus, it is helping to improve nature by doing what it "cannot" (see also Chávez-Courtois, 2004). For the paranatural discourse to work, nature and technology need to be configured in a particular way. Let me unpack this:

First, nature needs to be framed as not successful 100% of the times; for example, by presenting non-assisted reproduction (or spontaneous, as physicians call it) as essentially with low success rate: "as human beings we are not very good" when it comes to procreation, "every try holds only a 23% chance". Under this parameter, "even healthy fertile women can fall short of the standard" (Albury, 1999: 45). Then, by highlighting the aid technology can offer in "everything that nature can no longer do" and showing how nature can be improved with human technological assistance (see also Franklin, 2014; Franklin, Lury, & Stacey, 2000): "there is no reason why all the people here could not go home with a baby, everything has a solution" (Physician at Recruiting Event, 2009). Third, ARTs need to be framed as successful precisely because they "imitate nature" and because they substitute "the biological processes that originally take place in the organism [...] and that now can take place outside the body in lab conditions that faithfully reproduce the organic environment". In this way, ARTs imitates and goes beyond nature by using sophisticated technology that grants the possibility of achieving pregnancy and having a better offspring (without genetic alterations or of the desired sex). The final moves are stressing that assisted reproduction is scientific, "it is not magic". The sum of these qualities is what I call a *paranatural* framing: ARTs go parallel to nature yet they also reach beyond it. The idea of *paranatural* is an example of Franklin's notion of the "analogic turn". This notion "involves not only the 'borrowing' of analogies in one direction ('just like nature'), but also their ability to 'travel back' ('just like technology')" (Franklin, 2014). By performing this "analogic turn", the framing of assisted reproduction as a *paranatural* act becomes an entanglement of nature-technology, of science-culture, and of the sacred-scientific, an entanglement that makes assisted reproduction usable.

In the early 2000s, some clinics were already stating that, at their clinic, 96% of the couples achieve pregnancy; today (2019) they are now claiming 100% success rate or your money back. This "or-your-money-back-guarantee" rhetoric began with phrasings such as: "I assure you that you will have your baby". Why shouldn't they assure you if they depict science as capable of "defeating infertility".

This message was also narrated visually (Stephens & Ruivenkamp, 2016) at the information events, recruiting events, at the trade show, and in the media, through images and live interactions where what you see is what matters. This visual narrative was composed of: (i) technological portraits, which were blown-up images or presence of precision technology (microscopes, syringes, pipettes), digital equipment (incubators and freezers), and manipulated technological images of ova, sperm, and embryos; (ii) scientific portraits, such as charts and tables indicating numbers and percentages that reinforced what they were saying and suggested a high degree of predictability of the results; and (iii) family portraits, pictures of happy families, baby stuff (clothes, toys, etc.), and pregnant women or babies carried by their parents giving the marketing discourse an emotional touch and a promise of success. At the recruiting events and at the trade show, these family portraits made physical presence. Couples would come with their babies to talk about their success stories to current and future AR users. This materialised the possibilities promised by the doctors and their high-technology procedures. During the third year of Expo Fertilidad, a group of former AR users gathered, with their children, around the expo's organiser to have a group picture taken. The community of AR users gathered together for this picture were material proof of all the success stories of assisted reproduction.

Assisted Reproduction Is a Motherly Act

Within Mexican culture, a good mother, and hence a good woman, should follow the gendered guidelines of Marianism. These guidelines dictate that proper wives must be virgin before marriage, fertile, with motherly aptitudes, and healing capacities. They should be strong, willing to sacrifice everything for their family, abnegated, and subordinated to the male figures in her life (Napolitano, 2009; Pastor, 2010; Tate, 2007). Therefore, as wives, women are expected to serve and take care of their husbands and to "give" them children, and as mothers, they are expected to do everything and sacrifice all for their children. This dual role defines women's identity and purpose in terms of her husband and her child, which united constitute her family. A woman with a family becomes an "ama de casa" (housewife) with a purpose in life: take care of her family (Asakura, 2005; Neyer & Bernardi, 2013; Tate, 2007). In the following two sections, I analyse two instances where this is very clear: the interactions held in AMI's online forum and the telenovela La Rosa de Guadalupe (La Rosa). Both of these instances are examples of how assisted reproduction is framed as a motherly act making it culturally acceptable and usable. What I found in the forum and in La Rosa was present also in advertisements and articles published in magazines and newspapers.

AMIs, mAMIs, MB, and Other Forms of Motherhood

People who turn to assisted reproduction as a way of reproduction and family building commonly end up in what I have called the AR pilgrimage (2016) and what Franklin (1997) called the obstacle course. In the context of the early days of IVF in Britain, when people's life-course was interrupted by infertility, conception was no longer a taken for granted event but had to be achieved, and in order to achieve it, there was a "sequence of 'hurdles' to overcome" (Franklin, 1997: 144). In this context, Franklin heard people undergoing IVF speak of the "obstacle course": a "metaphor used to present the various stages of IVF and the physical, emotional and psychological difficulties they pose" (Franklin, 1997: 102). The people undergoing IVF Franklin spoke to, would described IVF as a procedure or experience that was "taking over" their life, eventually becoming a way of life. In a similar way, the people I spoke to during my fieldwork embarked in a pilgrimage, a sort of journey through clinics, doctors, attempts, stages, emotions, etc. This pilgrimage began when people decided to have children and became concerned because they felt they were being unable, it included visiting many doctors, trying many different options, and has very different outcomes. As the following quotes indicated, at the core of the pilgrimage is the jumping between physicians, clinics, and procedures due to the uncertainty, the desperation, the hope, and the despair that envelops the process:

> When you find out that someone got pregnant with one particular doctor, you just want to run to him. (Female, more than 6 years of AR treatments)
>
> I change doctor every time the treatment fails. (Female, 13 years of AR treatments)
>
> We change clinics when we have too many cycles or when someone recommends us a new one. (Female, 19 years of AR treatments)

The sum of these experiences makes up their AR biography: their diagnosis, the years they have been trying to conceive, the number of doctors, clinics, and treatments they have tried, the experiences they have accumulated, the failed attempts, and all that has happened in their journey to try to become parents.[19]

---

[19]All the interactions I read on this and other forums, as well as the interviews I held with patients at clinics and information sessions, were all "in real time". This means they were, at the moment of my encounter with them, undergoing a cycle. Recently, I have interviewed former users, and the way they speak of this pilgrimage is different. Further research on the difference between these narratives would be interesting.

Within the AMI community, members assembled visual and textual elements to create an avatar or portrait of their AR biography, of
their own individual reprohistory. This portrait depicted that they were
involuntarily childless, that they were users of ARTs, and members of
the AMI community. AMIs,[20] as they would call themselves within the
forum, presented themselves as not simply women who have not been
able to have children; they portrayed themselves as women who were
doing something, doing everything they could, to solve this problem. They were actively trying to become mothers, and in a way, they
were becoming good mothers in doing so (we could even argue that
they were already mothers).

This particular forum included two stages: undergoing treatment,
which they called Projecto BB (BB Project), or already pregnant, which
they called mAMIs (mAMIs means mommies in Spanish). If motherhood was achieved via adoption, they became mothers of a different
order than those going through pregnancy. These women were called
"Mamás de Corazón" (Heart Moms), their child *"Angelito de Corazón"*
(Angel of the Heart), and the biological mothers were nicknamed *"MB"*
(which stands for biological mother). Their portrait showed the stage
of the process they were at, the type of procedure they were using, and
the degree of participation within the AMI community (different status
were given to members depending on their involvement: distinguished,
platinum, silver, golden, and brass members). Looking at their portraits,
you could read their AR biography and appreciate the long pilgrimage
and difficult obstacle course these people had undergone. Their portrait had (again following the logic of Arcimboldo): their online name,
which could be a pseudonym; an image which sometimes was a picture
of themselves or of their child, an image of a baby, or of an angel; their
member status with words and ribbons; where they lived; their medical diagnosis; the type and number of previous treatments as well as
their outcomes; the number of lost pregnancies or miscarriages; the age
of their living children; their due date of birth or arrival (in the case

---

[20]AMIs sound similar to the word for friends in Spanish "amigas" and particularly to an endear-
ing slang "amiguis". With this term, they position themselves at a fraternal level.

of adoption); a thought; and a link to their blog or YouTube video. This portrait depicts the amount of engagement and scarifies these women have undergone for their child-to-be. This portrait works as proof of everything the woman has done to become a mother. But not any kind of mother, this portrait also works as a platform to position themselves as good marianistc mothers: a mother that sacrifices herself gives everything and more, for her children, even before this child is physically present.

Surrogacy as a Motherly Act

The opening scene of La Rosa's *A Light of Hope* is at Roberta and Santiago's civil wedding where Roberta's mother and Santiago's father lay out the rights and responsibilities she now has as a married woman. Her mother says: "You deserve to be happy. You have gotten married in the Church and now you are legally married, I wish you happiness", and her father-in-law says: "Now that you are the wife of my only child, I hope that soon you will give me many grandsons" (Next scene). Three years after the wedding, Roberta and Santiago are at a weekend family luncheon. Roberta's mother stresses the importance of family for women's purpose and identity; she says: "My family is my reason for living". Listening to this, Roberta gets up and goes to wash dishes, her mother—displaying her motherly sixth sense—detects that something is not right and goes after Roberta and finds her holding back the tears. She breaks down: "It's been three years and I haven't been able *to give* Santiago a child. I'm afraid that something is wrong" (emphasis added). Roberta's mother comforts her and tells her she needs to look for professional help and that she, as her mother, will help her and will always be there for her. Roberta starts visiting doctors and begins treatment. Throughout the following scenes, Roberta constantly, dramatically, and always in tears states how unhappy, unworthy, useless, empty, and unwomanly she is because she cannot "give" Santiago a child, and her mother firmly, passionately, and actively insists they will find a way, they need to put up a fight, do everything to achieve motherhood.

"Don't feel defeated before putting up a fight. It won't be easy, but I will be next to you all the way. I would give anything to avoid your suffering, anything!". After repeated failed attempts, they think about adoption, but Santiago's father is categorically against it:

Scene 19: Roberta and her father-in-law

Santiago's father: You know I am a business man and I don't like to beat around the bushes. I am here to ask you to divorce my son. I know you can't have children, and a woman like that is no good in my family. During generations we have made a great fortune, and it must continue this way. That is why I need a heir. Love has nothing to do with this. The important thing here is procreation, to reproduce oneself through the children.
Roberta: There are other options, like adoption.
Santiago's father: Never! I will not allow a fortune of this size to end up in the hands of a stranger. My heir has to be blood of my blood. The only solution is that you get a divorce. I will guarantee that you will be well off financially.

Appalled by what he is suggesting, Roberta kicks him out of her house.

A few scenes ahead, Santiago is caught having an affair with his secretary, an affair that is presented as orchestrated by his father to lure him away from Roberta. Santiago and Roberta brake up, but after months of being unhappy, Santiago breaks ties with his father and goes back to Roberta. He begs her for forgiveness. They get back together. When they tell Roberta's family that they are getting back together, Roberta's younger sister asks if they will abandon the pursuit of a child. They answer that they will not, but that they will now try without the pressure of Santiago's father. Nonetheless, Roberta is still unhappy:

Scene 22: At Roberta's home

Roberta's mother has come back from finding a new treatment that seems to offer hope.

R: I have had so many failures, I have lost all hope.

M: Never surrender, while there is life there is hope to fight.

R: I no longer have strength to face another deception... I have so little strength I must use it to save my marriage... I suspect S is involved with someone else.

M: If that is what you think, now more than ever you have to fight to recover your husband. He wants a child, and you have to give it to him... there are no buts, fight, fight you have to fight until the end.

R: you are right mom, I will fight. We will go see the Dr.

Roberta's mother takes her to see the doctor to find out about the new treatment. The doctor says that she is a perfect candidate. They gather the family around to give them the news: they had found a treatment that would work. The treatment would be surrogacy, and Roberta's mother would be the surrogate. The family disapproves. Roberta's father and siblings are not happy with this. In one scene, Roberta's younger sister accuses Roberta of being selfish and putting their "mother on the line just so you can have children". This made Roberta think twice; however, the mother insists "this is something I want to do and that I would do for any of my children". But neither Roberta nor her mother can do this alone. They need the blessing of the Virgin Guadalupe. We see the white rose (which indicates that the Virgin has listened to the prayers and will intercede). The Virgin has witnessed Roberta and her mother fight all the battles, and she has seen them suffer and sacrificed everything they have, all in order to become a mother—the purpose of women, the sacred gift, and responsibility bestowed upon them by God. The Virgin has approved. Roberta's mother will be the "human incubator" (which is the term used in the programme for surrogacy).

(But this is a melodrama, something has to go wrong) Within a few days of the due date, Roberta's mother falls unconscious, fighting between life and death, her voice says "How important it is for a woman to give life", and she awakes victorious over her death and over her daughter's infertility, saving her daughter's marriage, giving her daughter a reason to live, restoring her womanness, and re-enforcing her own:

Scene 36: When Roberta's mother is handing over the baby to her

R: Thank you mother for all the happiness you have given me, you gave
me strength when I had none.
M: Now that you are a mother you can understand that there is nothing
worst for a mother than to see your child's suffering.

The programme begins setting out the importance of marriage. Getting
married gives Roberta the right to be happy and the responsibility of
becoming a mother. Marriage contextualises and marks the beginning
of three new areas in women's life: her sexual life, her conjugal life, and
her reproductive life, and all are in turn supported by values of virginity, maternal love, and reproductive sex (Finkler, 1994; Sanchez Bringas,
2005). Only through marriage can women access the defining roles of
the Mexican female identity: the role of wife and the role of mother,
which conjoined become the role of housewife (Asakura, 2005). These
roles define women in terms of the other: their husband and their child,
which united, become her family. Therefore, women are women only
if they can produce offspring; otherwise, they become "empty and dry
women".

The next narrative point is showing Roberta's devastation for not being
a mother, for not being able to fulfil her womanly duty of *giving* her husband a child. The wife's responsibility towards her husband goes beyond
serving and taking care of him, she must also "give him a child". Several
scenes are filled with lines where Roberta is battering herself because she
"can't be happy", where she claims that the only way of being happy is
being a mother and "I will not be a mother, I will never be able to *give*
Santiago a child" (emphasis added). Without a child, how can she prove
she is a good wife, a good mother, and hence a good woman? Some
women feel that without a child, they "can't be happy". According to
them, the only way of being happy is being a mother because their family
is their "reason for living". Without a child, some women fee they lose
their value as women, their purpose in life, and fear they might lose their
husband, even their family (cf. Carreño-Meléndez, Morales-Carmona,
Sánchez-Bravo, González-Campillo, & Martínez-Ramírez, 2003).

A woman's responsibility as a mother is to do anything and sacrifice everything for her children; the more a mother sacrifices, the stronger the evidence of her maternal love, and the better mother/woman she is believed to be. Sacrifice is at the crux of these two relationships. Women must sacrifice everything for their husbands, and above all, for their children. Sometimes, they think they should sacrifice themselves, push their husbands away, and convince them to leave them, so they can fulfil their desire of having children with another "healthy" woman (see *Marie Claire*, September 2006 and Tu Fertilidad 2008). Roberta and her mother portray marianism in the most literal way: they are the good, pure, Catholic female figure (i.e. mother) who is happy to sacrifice herself for her family. However, the act of serving the family is not passive. These female characters are the ones who make things happen. We see Roberta doing everything she can to bring to life a child proving that she has what it takes to be a good mother. We see Roberta's mother constantly on the move looking for the new doctor, the best clinic, and the latest treatments. We see her keeping Roberta going, insisting that there is always another option. And finally, we see her bringing the ultimate solution: surrogacy. For Roberta's mother, her suffering childless daughter represents a problem she needs to solve. It was her motherly duty to do everything she could and sacrifice everything she had in order to help her. And that is what she did. She offered to be her daughter's surrogate.

The programme presents surrogacy as a viable option to build a family and as one that allows women to fulfil their marianistic role, since it implies giving one's husband a child and sacrificing everything for it: Roberta is scarifying her mother's life against her brother and sister's will, and Roberta's mother is scarifying her own life against her husband and her own children's opinions. Roberta's mother proves to be a great mother, as does the Virgin Guadalupe. All mothers unite to bear this child. In this telenovela, maternity is depicted as powerfully inspiring, it makes things happen, but it is also powerfully demanding, and it has to fulfil the demands the gender roles culture imposes.

In La Rosa, there is never a mention of a possibility of male infertility. In several scenes where Roberta is seen visiting clinics, doctors never mention the need to run tests on him, and none of the other characters ever suggests male infertility as a possibility. This reinforces the idea that reproduction is exclusively a female issue. In contrast to this narrative, in other media productions as well as in many of the information sessions organised by clinics, the incidence of male infertility is stated as being anywhere from almost equal to slightly above that of female infertility. As with female gender roles, in men the sexual and reproductive roles are also united to shape their gender identity. Sexual performance is associated with masculinity, virility, and power, which are displayed through the number of children he can produce (Asakura, 2005). If a man cannot have children, there is the belief that it is because he has a sexual dysfunction. This concept makes it difficult for men to accept being medically checked when male-related infertility is suspected.[21]

La Rosa shows another important side of adoption: the family's opinion matters. Many women and couples I spoke to at the AR clinics had thought about adoption, yet had decided against it for two main reasons: either they believed they were not eligible because of their income, their lack of patrimony, their marital status, or some other reason that made them think they would not be approved as a family worth of receiving a child, or their families had openly said they would not accept the adopted child as part of the family. Regarding the first reason, it became clear that the adoption process is full of myths and misguided information. Regarding the second reason, it became clear that, as opposed to some of the treatments in assisted reproduction, where users can withhold all or some information (e.g. the use of donated gametes), in the case of adoption this cannot be the case, so they need to consider their family's opinion and act considering the consequences of their opinions. Many were not prepared or willing to cut off ties with their families because of their rejection of adoption, so they had to try with ARTs. At the AMI, meetings and in the online forum adoption

---

[21]However, one of the few studies that looks at the attitude of Mexican men towards infertility and assisted reproduction stated that most of the male partners of women seeking AR treatment at that particular public institution were submissive, passive, and had not a strong desire to become fathers, but that they tolerated and collaborated with their wives during treatment (Arranz-Lara, 2001).

was presented as a valid way of becoming a family (as did the physicians at the AMEE).[22] In sum, this and other fictional stories as well as the ethnographic material I have collected show that the cost of remaining childless or of adopting can sometimes be higher than the cost of assisted reproduction.

Family and Gender

While at moments ARTs have been seen as granting the possibility and the opportunity to subvert gender roles, to rewrite them, this has not always been the case (De Saille, 2017). ARTs can equally be used to perpetuate heteronormative ideas about family and gender roles (as just described above in the case of La Rosa). In newspaper articles, for example, it was still common to find "the desire to become a mother" framed as a female instinct, which can be dormant during the stage some women want to pursue their career, but it will emerge when the biological clock's alarm is about to go off. Hence, female infertility is described as the result of postponing maternity—in turn framed as a voluntary action. Male infertility is presented as due to physical issues such as viruses, inflammation, or erectile dysfunction. However, men are not presented as having such a strong desire to become fathers. It is their responsibility, to perpetuate the family name and pass down

---

[22]"Everyone at AMI knows my situation and all the treatments that I have gone through…at the beginning I said NO TO ADOPTION…after six years of trying every treatment: IVF, sperm donations, embryo donation, herbal teas from the Juarez Market…we have considered renting a womb in the USA…you can't imagine what I would give to have a baby in my arms, to carry it, to kiss it. It doesn't matter if it is born out of me, if it has the sperm and ova of another couple, if the sperm is from a donor…come as it may, but I want to have a BABY IN MY ARMS and not necessarily in my womb…I think that MOTHER IS WHO RAISES YOU, WHO EDUCATES YOU, WHO LOVES YOU…and not necessarily who carries you in her womb… surrogate mothers are women who simply are the oven in which the cake is baked and you pay them to have them grow your baby in their womb…so by whichever method, the objective we all have is TO HAVE A BABY IN OUR ARMS TO KISS, TAKE CARE OF AND TO HAVE HIM UTTER THE WORD MOM. WHICHEVER PATH YOUR BABY TAKES TO ARRIVE IS NOT BETTER NOR WORST…OVA DONATION, SPERM DONATION, EMBRYO DONATION, INSEMINATION, IN VITRO, ADOPTION, SURROGATE…ETC…and yes, it will not have your eyes, but it will have your gaze" (Forum post, emphasis in the original).

the fortune (as was depicted in La Rosa) but not so much as source of identity or purpose in life. Mexican masculinity is more commonly built on notions of work, sexual potency, and caring for the family (in terms of physical and financial protection) (Wentzell, 2013). ARTs were described as the infallible solution to all these problems.[23]

Biomedical reproduction, both as family planning or as assisted reproduction, is framed as a modern way of attending a female matter (problem, desire) by male experts. This was clear, for example, in the selection of experts invited to the information sessions organised by AMI, to give the talks at Expo Fertilidad, and to present the procedures at the recruiting events. They were talking about infertility as a situation—mainly located in the female body—that can be solved using the proper procedures offered by them, almost all male gynaecologists, experts in the field of biomedical assisted reproduction. In the talks, gender roles were constantly re-enforced. For example, at the beginning of the information sessions, the presenter would thank men for being there, sometimes even using the phrase "accompanying their wives", as if reproduction were not an issue that occupied them, or in the covers of Tu Fertilidad and in the plot of La Rosa, where men's desire for children is not explored. Even in recent years, when AR clinics are now openly addressing same-sex couples, it is more frequent to find messages directed to women than men.

However, I do think it is important to contextualise all this. This book is tracing the conformation and development of the AR system. This also means tracing how things change. In this respect, the way family and gender are depicted, understood, and talked about within the system of assisted reproduction has also changed. It has done so slowly, yes, but at this stage, between 2000 and 2010–2012 these changes were detectable. For example, while at Expo Fertilidad the discourse was predominantly heteronormative, the specialised magazine, Tu Fertilidad, offered a more inclusive message. The magazine did include articles on single people (both men and women) and same-sex couples using ARTs

---

[23]See, for example, Aguilar, El Universal, 2000; Pérez Stadelman, El Universal, 2001. For a full list of articles, see González-Santos (2011).

as well as many articles on male infertility. Another example was the concept of family portrayed by AMI. While their slogan ('Formando Familias' forming families), their logo (a design of a sperm entering an ova), and the images on their website (a baby with its father and the face of a baby) all stress the idea that a couple becomes a family when there is a baby at the centre, during the events they offered another version of what is needed to constitute a family. In these talks, they emphasised that a couple without children is also a family: "infertility is a family issue, because the family is the couple; it is not true what the media says that in order to be a family you need children, because a family can be just two" (speaker at AMI information event, 2007).

"Artefacts have politics", this premise is widely accepted within STS (Winner, 1980). So the question is, how do these politics clash, dialogue, or morph when faced with the politics of a given culture?[24] In the case of Mexico, there is one aspect of assisted reproduction that very clearly dialogues with Mexican culture: the idea that having children is important. Having children is central to Mexican female gender identity, and it is a vital element for survival within Mexican society, a nation that strongly relies on the family, not only when facing a crisis but in general, for everyday life matters. Precisely because family is the only relatively trustworthy institution, being blood related is still considered very important. However, assisted reproduction has dismembered the concept of "blood related" in at least two kinds of links: genetic and gestational. Although participants in this study presented perspectives that varied considerably on which of these two links is more important to them, it was evident that these were perceived as distinct. The current law favours the gestational link over the genetic one.

Assisted reproduction has widened the range of biologically linked family systems (see the recently popular procedure within Spanish speaking countries called ROPA, which stands for reception of partner's

---

[24]This is one of the central questions within the field of AR studies. For a group of researchers concerned with this, see, for example, the group Changing (In)Fertilities (Reprosoc Cambridge) and AFIN (Universitat Autònoma de Barcelona).

eggs, recepción del óvulo de la pareja). Within Mexico, the use of ARTs to promote the traditional heterosexual family system has been well received, less but still somewhat acceptable is the use of ARTs to help single women fulfil their role as mothers. This points out another strong element that bonds assisted reproduction and Mexican culture: not only does assisted reproduction allow for motherhood to become a reality and not only does it reinforce the idea that every woman's dream is to become a mother, that every couple's purpose is to become a family, and that a family is conformed of a heterosexual couple and children, but above all, it highlights the notion that women should sacrifice everything they have for their children (or children to be) and that science—aided by God—is capable of everything. Assisted reproduction strengthens one more very important element in the Mexican construction of motherhood, the element of sacrifice. Assisted reproduction is a sacrifice that submitting to it is considered a motherly act, allowing women to become good mothers before the child is even a material reality.

The way assisted reproduction has been adopted by Mexican society has very strong practical (pragmatic) side. Although I only spoke to those who had already decided they would use ARTs (and regardless of how that decision was taken in terms of the information they had received), I was able to observe the decision process when a new procedure, usually of higher complexity, was being suggested to them. In this process, the aspects that played an important part in the decision making were a combination of: economic costs, understanding the diagnosis and the procedures, fears related to the acceptance of the child (with gamete and embryo donation people wondered: will I love it as much, will it look like us, will people know). Many Mexican users articulated technology, culture, and even religion in a way that helped them make sense of and accept assisted reproduction. In the previous pages, I have detailed two ways in which assisted reproduction is made usable: first by framing ARTs as paranatural technologies and second by framing their use as a motherly act. In the following section, I discuss the commercial aspect of the AR system.

The Commercial Touch

I think there is a boom in infertility, but I am not sure it is for the best. There are positive aspects, like for example the prices have gone down, there is more and more accessible information. But the negative aspects of the boom is that it has become a quick and apparently easy solution for infertility problems and this leads doctors to under-diagnose and under-explore. [...] the tendency is tilting towards skipping the simple procedures and going directly to IVF. [...] Clinics are now advertising in the bus stops and billboards. In the past, if a doctor advertised, he was frowned upon.[...] as I say, I am not sure to what extent we are doing things right or wrong, what I can say is that all this has acquired a commercial touch to it, but it works. People are overcoming their infertility and they are getting pregnant. [...] At the end, I think AR is working but I am not sure it is heading in the right direction. There is a boom, but like the virus, they will self limit their growth. Some clinics will close, because although I do think the cake has gotten bigger, it has not duplicated or triplicated. The infertility patients are still the same, maybe 10% more. I think that at some point this will implode, some will have to close and my instinct says it will be most likely the smaller ones. (Marketing Expert, Interview 2008)

The above excerpt is taken from an interview I had with a marketing expert who had been involved with the promotion of AR clinics, pharmaceutical companies, and with informing patients. This interview, as the quote proves, was very revealing in terms of what was happening during the first decade of the twenty-first century. There are things I want to highlight from this quote. First, there was a boom in the industry, but booms can implode, and they can only go so far. Second, booms can bring positive aspects, like prices going down, but also negative ones, like under-diagnosis. Third, ARTs work. While keeping its framing as a very specialised and high-tech set of procedures capable of producing mothers (as discussed above), assisted reproduction also became a way of commodifying infertility and reproduction. All this is part of the "commercial touch" my interviewee identified. This touch, I argue, has helped expand the universe of assisted reproduction, has made it usable, has contributed to changing how we see kinship, reproduction, and the interaction between nature/culture and technoscience, and has been the result and the promoter of the establishment of a profitable market. Let me describe this in a bit more detail.

The boom that this market expert talks about refers to the expansion of the universe of assisted reproduction (see Chapter 5). Between 2000 and 2010, there was an increase in the number of clinics, these were now located in more places across the country, there were more specialists graduating from the specialty, there were more procedures and the patient eligibility for these procedures expanded, and thus, there were also—apparently—more potential users. But offer and demand within the private sector were not growing at the same rate, and so competition began, particularly where there was a higher concentration of clinics. The "cake" (i.e. patients in need of assisted reproduction), as said in the quote, had to be split between more clinics. As the universe of assisted reproduction expanded, and there were more competing clinics, advertisement increased.

Clinicians and clinics first used the hybrid articles mentioned above, part science communication part publicity, as their promotional tool. They also set up websites. In the beginning, these websites housed mostly practical information that was presented as such: office hours, location, list of services offered, name of service providers, credentials, accreditations, and certifications. Soon, the way this content was presented changed, and it began to include other sorts of visual and textual messages. It was not just services, now the idea that their clinics were professional spaces with first class high-tech, depicting this through images of the clinic's labs, and the high-tech tools. The variety of services that they offered in addition to IVF (what have been called the add-ons) such as gamete donation, PGD, sex selection, or even surrogacy were highlighted. Some clinics also resorted to infomercials and franc advertising in the press and the media. In doing this, contemporary advertising (which now includes the use of social media) has contributed to overflow the biomedical framing of assisted reproduction to include new users and new uses (Çalışkan & Callon, 2009; Dyer, 1997). More clinics began to engage in publicity campaigns where they offered discounts and reduced price packages. Although "persuasive advertising has been considered ethically dubious because it exploits hopes, fears, and desires of potential consumers" (Crisp, 1987: 196), today (in 2018) it is being practised by several clinics.

The increase in number of clinics and their wider distribution across the country increased the availability and accessibility of AR services, and, according to the marketing expert in the quote, competition could have contributed to a stabilisation of prices. The cost of AR services is determined by a series of factors, some quite predictable, like the doctor's fees or the cost of specific procedures such as blood tests, ultrasounds, egg retrieval, or AI. There are, however, other things that are more unpredictable, like the cost and amount of drugs required for the ovarian stimulation, or the number of eggs and embryos that will be viable, and if there will be a need for freezing some embryos. So "saving up for a cycle" is not like saving up for a car. The cost might vary considerably as the cycle proceeds to the point that, in some cases, patient's had to abandon the cycle without having finished due to a lack of funds or resources. Depending on what sort of healthcare system the person is aligned with, he or she will face different costs and limitations to the AR service. While most public healthcare systems (IMSS, ISSSTE, PEMEX, and SS) offered some sort of AR services, they do so with limitations, not only in the procedures they offer, but also in terms of the eligibility of the patient. And in both the private clinics and the SS, there is still a cost that needs to be covered out of pocket.

Until recently (2018), no insurance company operating in Mexico covered procedures for assisted reproduction. Not only that, but they would also limit the coverage of children conceived through assisted reproduction. Given this context, it is not surprising that there is a large sector of users who pay for these services out of pocket or through loans. During the first decade of 2000, there was one Mexican bank—IXE—who offered a financing scheme called Matices (which in Spanish means shades or tones) which was specifically designed to help cover the cost of medical interventions intended for the improvement of one's "personal image" and included treatments such as bariatric and plastic surgery, dermatology, gynaecology, hair implants, nutrition, odontology, ophthalmology, otorhinolaryngology, dental treatments, and fertility procedures.

Within the context described in the quote where the "cake" was not getting bigger and yet the number of clinics was increasing, it became important, particularly for new clinics, to find ways of attracting those potential users who were being left out due to the high costs of the treatments. Some clinics began to offer pre-paid packages of 3 or 4 cycles, others to offer discount, or to raffle full cycles at the trade shows or at the recruiting events. During the first year of the trade show, most clinics offered 50% discount in the first consultation and some raffled full free AR cycles. Then during the second year, one clinic offered a fixed price for all the procedures needed to get the person pregnant and, at the third year, another clinic offered an insurance policy: if the first IVF cycle was unsuccessful, the clinic would pay for the second cycle. Here, we see a transition from only 50% discount, to fixed price, to a sort of guarantee, and even a refund in cases of an unsuccessful attempt. After 2010, this guarantee turned into a package-guarantee scheme where people pay for three cycles and if they don't get pregnant they get their money back. These strategies could seem like a way of saving money, but this rarely occurs since there were always hidden charges or unexpected costs.

A peek into the future. In 2018, most clinics have Facebook pages, in them they offer discounts on specific dates. For example, during February due to Valentine's day, April due to Children's day, May due to Mother's day, June due to Father's day, December due to Christmas, and some during the clinic's anniversary. Most clinics also adhered to the Buen Fin campaign which is inspired on the American Black Friday.[25] This illustrates how the AR system has participated in the marketisation of the health business in Mexico and how reproduction has been commodified. This has happened as part of Mexico's bioeconomy project.

---

[25]While the Black Friday was thought of as a strategy to get rid of the last season's goods and make way for the new ones that will be feeding the consumers for Christmas, the Buen Fin is simply a way to boost consumption. It is a annual nationwide shopping weekend scheme implemented since 2011 to encourage people to spend money by offering special promotions (e.g. extended credit terms, points schemes, and store credit). It takes place during the weekend of the Mexican Revolution holiday (20 November). Critics say this only encourages unnecessary consumption leading to debts since buyers tend to pay with credit cards in monthly interest-free regular payments. It conjoins the interests of the Mexican Banks association, the Mexican Internet association, the Commerce and Tourism Chamber (CONCANACO), the Industry Chambers (CONACAMIN), and the Mexican government.

During the past fifty years, Mexico has increasingly participated in a medical bioeconomy project. This project deploys a narrative articulated through a series of meta-narratives such as the human rights discourse, the centrality of the individual, the success of neoliberalism, and the hopes and hypes around biotechnology. The articulation of these narratives has contributed to the growth of the AR market. These narratives, together with cultural narratives concerning gender roles, fruitfully articulate to make assisted reproduction useable. This bioeconomy project has also participated in the medical tourism industry. This industry has been fulled by several actors and situations. As far back as the 1960s, hispanic workers living in the border cities in the USA would come to Mexico in search of dental care. Since the 1990s, there has been a growing population of US citizens moving to Mexico to retire and seeking more affordable health care. More recently, with the availability of ARTs at lower prices and with comparable success rates, and during a period when surrogacy was accessible, there were many foreign individuals that would come to Mexico in search of AR services. Nuevo León was a pioneer state to formalise the medical tourism industry with the association of Christus Health Foundation from Texas and the Muguerza Group in Monterrey and then through the creation of the CIMA, established by the International Hospital Corporation Group. These two groups cared for patients coming from the USA, Tamaulipas, Chihuahua, and Sonora. This medical tourism industry has affected the way medicine is now being practised in Mexico. For example, it has lead to alliances between hospitals and the hotel industry, and some hospitals have sought quality certification standards that meet the USA and Canadian criteria as part of the North America Free Trade Agreement (NAFTA) (e.g. The National Certification System of Medical Care Establishments, SiNaCEAM). Overall, the neoliberal project has affected the relationship between physicians and hospitals. It is now becoming common practice to pressure physicians to have their patients use the hospital's installations and services and to promote hospitalisations. In the case of assisted reproduction, some practitioners have suggested this discourages single embryo transfer, since a multiple pregnancy is always a profitable event for the hospital.

Final Thoughts

Some final thoughts about what has been discussed thus far. The focus of attention in this chapter was the way assisted reproduction was depicted in the discursive landscape, an environment where visual and textual elements collectively and collaboratively produced a complex narrative on reproduction, kinship, science, nature, culture, and more. I saw how, in this complex narrative, assisted reproduction was depicted as a paranatural technology capable of producing marianistic mothers. Equally present in this landscape, was evidence of the process through which reproduction, infertility, and assisted reproduction were commodified and marketised.

Commodification, in the narrow sense, means that something can be bought and sold. In a more broad sense, however, it means that the "something" is being regarded in terms of a market rhetoric: thinking about, talking about, and interacting with the "something" as if these interactions were sales transactions. This includes "owing, pricing, selling, and evaluating interactions in terms of monetary cost-benefits analysis" (Radin, 1987). In this chapter, in dialogue with the rest of the book, I looked at how the commodification process was precisely that, at situated process. This commodification process was the result of, and not the instigator of, the different elements that make up the contemporary AR system and that were highlighted in this analysis of the discursive landscape.

The media and the clinics found a profitable alliance in this landscape: stories of how people achieved pregnancy through the use of these spectacular emerging biotechnologies made news-worthy stories; physicians had the opportunity of broadcasting their name and advertising their clinic; and potential users found in these spaces stories about how assisted reproduction can be used, by showing, the instances, purposes, subjects, motives for when to use ARTs. Advertising is a key feature of the market economy. It is a way of making a good or service conceptually accessible, in terms of bringing the consumer and the product together, of presenting the service so the potential consumer knows it exists, and showing the different ways it can be consumed. The particular persuasive way of thinking about, speaking of, and interacting with the "thing" that is being advertised in order to justify its existence and use (Phelan, 1996) frequently juxtaposes potential future outcomes (you will get pregnant) to communicate support for a given action (the use of assisted reproduction). They bring in past examples from other cases to predict the success in this current case and thus promote their use.

Commodification is a situated process. In this regard, I argue commodification of assisted reproduction was possible in this case because it was depicted as a successful and low-risk procedure capable of offering people what they wanted: to become a (good marianistic) mother. Furthermore, it is technologically flexible enough to accommodate a variety of users and uses. What is now happening is the sociocultural and legal negotiations to allow for these new users and uses to become normalised.

When talking about the marketisation of assisted reproduction, my interviewee spoke about under-diagnosis. This brings up many interesting points. First, she was pointing out that it was becoming a common practice for AR physicians to receive patients and more quickly than before en-role them in the IVF protocol. It would be interesting to see if, in fact, there has been an increase in the percentage of patients going into AR clinics are being enrolled in IVF and ICSI procedures. Second, a practice of under-diagnosis could also be because the sort of people seeking help has begun to change or, I should say, a new group of people are also seeking these services (this goes along the lines mentioned above of the new users and new uses).

In the beginning, IVF was considered a procedure only adequate for specific cases of infertility. Now IVF is being used for other reasons as well, many of which are not necessarily always seen as medical. Under a medical framework, a single person or a same-sex couple are not, per se, infertile, nor are people wanting to balance the sex-ratio of their families. However, these are new uses for ARTs. And on many occasions, the legal and ethical debates stagnate because they are not highlighting that these are two different sets of users and uses. I call for further research on this point. I wonder if two approaches to assisted reproduction are developing, one restorative and one re-configurative (cf. Hables Gray, 2002). While the restorative would be faithfully reproducing the long existing, normative, kinship system, the re-configurative could be subverting this system, proposing and experimenting with new ways for forming and being a family. This re-configurative would allow for a new definition of eligibility to these services (moving away from the definition of infertility which presupposes a heterosexual stable couple when it requires twelve months of unprotected frequent sexual intercourse). Following what is now a common terminology within STS, this can be thought of as an interpretative flexibility where one or several new social groups are becoming relevant—same-sex couples, singles, people wanting to chose certain characteristics for their children—and thus changing the way ARTs are understood: no longer only as a procedure used to cure a disease but also as a procedure that can be used to achieve a desired goal, children. In other words, this ARTs are not only responding to, but also actively participating in changing perceptions and definitions of fertility and reproduction, the culturally established reproductive patterns and trends, regimes of who can choose how and when to reproduce, new kinship ties. This book fits into the growing number of scholars and research projects concerned with what is being referred to as studies in "Changing-In/Fertilities" (Franklin and Inhorn, Reprosoc).

References

Abusief, M., Hornstein, M., & Jain, T. (2007). Assessment of United States fertility clinic websites according to the American Society for Reproductive Medicine (ASRM)/Society for Assisted Reproductive Technology (SART) guidelines. *Fertility and Sterility, 87*(1), 88–92. https://doi.org/10.1016/j.fertnstert.2006.05.073.

Ahedo, A. (2017, December 12). Más de 7 millones de personas visitan la Basílica de Guadalupe. *El Universal*. Retrieved from https://www.eluniversal.com.mx/metropoli/cdmx/mas-de-7-millones-de-personas-visitan-la-basilica-de-guadalupe.

Albury, R. (1999). *The politics of reproduction: Beyond the slogans*. Crows Nest, NSW: Allen & Unwin.

Arranz-Lara, L. (2001). El deseo de maternidad en pacientes sujetas a tratamientos de reproducción asistida: una propuesta de psicoterapia. *Perinatol Reprod Hum, 15*(2), 6.

Asakura, H. (2005). Cambios en significados de la maternidad: la emergencia de nuevas identidades femeninas. Un estudio de caso: mujeres profesionistas de clase media en la Ciudad de México. In M. Torres (Ed.), *Nuevas maternidades y derechos reproductivos* (pp. 33–59). México: El Colegio de México.

Barbey, C. (2017). Evidence of biased advertising in the case of social egg freezing. *The New Bioethics, 23*(3), 195–209. https://doi.org/10.1080/2050 2877.2017.1396033.

Basten, S. (2010). Television and fertility. *Finnish Yearbook of Population Research, XLV*, 67–82.

Beard, L. J. (2003). Whose life in the mirror?: Examining three Mexican telenovelas as cultural and commercial products. *Studies in Latin American Popular Culture, 22*, 73–88.

Becker, G., Butler, A., & Nachtigall, R. D. (2005). Redemblance talk: A challenge from parents whose children were conceived with donor gametes in the US. *Social Science and Medicine, 61*, 1300–1309.

Bharadwaj, A. (2000). How some Indian baby makers are made: Media narratives and assisted conception in India. *Anthropology & Medicine, 7*(1), 63–78.

Çalışkan, K., & Callon, M. (2009). Economication, part 1: Shifting attention from the economy towards processes of economization. *Economy and Society, 38*(3), 369–398. https://doi.org/10.1080/03085140903020580.

Carreño-Meléndez, J., Morales-Carmona, F., Sánchez-Bravo, C., González-Campillo, G., & Martínez-Ramírez, S. (2003). En parejas estériles por factor masculino y femenino. *Perinatol Reprod Hum, 17*(2), 11.

Cerón, R. (2007, March 1). La ciencia vence a la infertilidad. *El Universal.* Section: Salud.

Chávez-Courtois, M. L. (2004). Presencia de esterilidad: Actores o sujetos en la actualidad. *Escuela Nacional de Antropología e Historia (ENAH). Cuicuilco, 11*(31), 1–17.

Crisp, R. (1987). Persuasive advertising, autonomy, and the creation of desire. *Journal of Business Ethics, 6*(5), 413–418. https://doi.org/10.1007/ BF00382898.

De Saille, S. (2017). *Knowledge as resistance: The feminist international network of resistance to reproductive and genetic engineering*. London: Palgrave Macmillan (Imprint).

Dyer, A. R. (1997). Ethics, advertising, and assisted reproduction: The goals and methods of advertising. *Women's Health Issues, 7*(3), 143–148.

Finkler, K. (1994). *Women in pain: Gender and morbidity in México*. Philadelphia: University of Pennsylvania Press.

Flores Palacios, M. L., & Sánchez Santana, A. G. (2006). Capítulo 11. Estereotipos demográficos y ocupacionales de la mujer y el hombre en la televisión mexicana. In *Anuario de Investigación de la Comunicación CONEICC XIII* (pp. 257–271). Retrieved from https://www.researchgate. net/profile/Ana_Gaspar4/publication/27392963_Rehacer_el_tejido_de_ Penelope_mujeres_y_reproduccion_de_la_emigracion/links/53ee50850c- f23733e80c528e.pdf#page=257.

Franklin, S. (1997). *Embodied progress: A cultural account of assisted conception.* London: Routledge.

Franklin, S. (2014). Analogic return: The reproductive life of conceptuality. *Theory, Culture & Society, 31*(2–3), 243–261. https://doi.org/10.1177/ 0263276413510953.

Franklin, S., Lury, C., & Stacey, J. (2000). *Global nature, global culture.* Sage: London.

González-Santos, S. P. (2011). *The sociocultural aspects of assisted reproduction in Mexico* (Doctoral, University of Sussex). Retrieved from http://sro.sussex. ac.uk/7081/.

Hables Gray, C. (2002). *Cyborg citizen.* New York and London: Routledge.

Hawkins, J. (2013). Selling art: An empirical assessment of advertising on fertility clinics' websites. *Indiana Law Journal, 88*(4), 1148–1179.

Hawkins, J. (2017). Exploiting advertising. *Law & Contemporary Problems, 80*(3), 43–71.

Lewkowicz, E. (2015). Cinderella's having a ball: Humoring Mexico's "ugly" TV formula. *Feminist Media Studies, 15*(2), 271–286. https://doi.org/10.10 80/14680777.2014.924980.

Madeira, J. L. (2013). Selling art or selling out?: A response to selling art— An empirical assessment of advertising on fertility clinics' websites. *Indiana Law Journal, 88*(4), 1181–1185.

Maher, J. (2014). Something else besides a father: Reproductive technology in recent Hollywood film. *Feminist Media Studies, 14*(5), 853–867. https:// doi.org/10.1080/14680777.2013.831369.

Michelle, C. (2007). "Human clones talk about their lives": Media representations of assisted reproductive and biogenetic technologies. *Media, Culture & Society, 29*(4), 639–663. https://doi.org/10.1177/0163443707078425.

Mol, A. (2002). *The body multiple: Ontology in medical practice.* Durham: Duke University Press.

Mulkay, M. (1994a). Science and family in the great embryo debate. *Sociology, 28*(3), 699–715. https://doi.org/10.1177/0038038594028003004.

Mulkay, M. (1994b). The Triumph of the pre-embryo: Interpretations of the human embryo in parliamentary debate over embryo research. *Social Studies of Science, 24*(4), 611–639.

Napolitano, V. (2009). The Virgin of Guadalupe: A nexus of affect. La Virgen de Guadalupe: Un Point de Fusion Des Affects. *Journal of Royal Anthropological Institute, 15*(1), 96–112. https://doi.org/10.1111/j.1467-9655.2008.01532.x.

Nerlich, B., Johnson, S., & Clarke, D. D. (2003). The first 'designer baby': The role of narratives, cliche's and metaphors in the year 2000 media debate. *Science as Culture, 12*(4), 471–498. https://doi.org/10.1080/09505 43032000150328.

Neyer, G., & Bernardi, L. (2013, August). *Feminist perspectives on motherhood and assisted reproduction.* Paper presented at the XXVII International IUSSP Conference, Busan, Korea. Retrieved from http://iussp.org/sites/default/files/event_call_for_papers/IUSSP%20Neyer%20Bernadi%20-%20 Reproduction%20updated.pdf.

Orozco Gómez, G. (2006). La telenovela en México: ¿de una expresión cultural a un simple producto para la mercadotecnia? *Comunicación y Sociedad* (6), 11–35 (ISSN 0188-252X).

Osborne-Thompson, H. (2014). Seriality and assisted reproductive technologies in celebrity reality television. *Feminist Media Studies, 14*(5), 877–880. https://doi.org/10.1080/14680777.2014.952876.

Oudshoorn, N., & Pinch, T. (Eds.). (2003). *How users matter: The co-construction of users and technologies.* Cambridge: MIT Press.

Pastor, M. (2010). El marianismo en México: una mirada a su larga duración. *Cuicuilco, 17*(48), 257–277.

Phelan, J. (1996). *Narrative as rhetoric: Technique, audiences, ethics, ideology.* Columbus: Ohio State University Press.

Radin, M. J. (1987). Market-inalienability. *Harvard Law Review, 100*(8), 1849. https://doi.org/10.2307/1341192.

Roberts, E. F. S. (2006). God's laboratory: Religious rationalities and modernity in Ecuadorian in vitro fertilization. *Culture, Medicine and Psychiatry, 30*(4), 507–536. https://doi.org/10.1007/s11013-006-9037-8.

Roberts, E. F. S. (2007). Extra embryos: The ethics of cryopreservation in Ecuador and elsewhere. *American Ethnologist, 34*(1), 181–199.

Roberts, E. F. S. (2016). Resources and race: Assisted reproduction in Ecuador. *Reproductive Biomedicine & Society Online, 2,* 47–53. https://doi.org/10.1016/j.rbms.2016.06.003.

Sánchez-Bringas, Á. (2005). Prácticas reproductivas en el Distrito Federal a finales del siglo XX. En M. Torres (Ed.), *Nuevas maternidades y derechos reproductivos* (pp. 33–59). México: Colegio de México.

Shalev, S., & Lemish, D. (2012). Dynamic infertility. *Feminist Media Studies, 12*(3), 371–388. https://doi.org/10.1080/14680777.2011.615627.

Stephens, N., & Ruivenkamp, M. (2016). Promise and ontological ambiguity in the in vitro meat imagescape: From laboratory myotubes to the cultured burger. *Science as Culture, 25*(3), 327–355. https://doi.org/10.1080/095054 31.2016.1171836.

Tate, J. (2007). The good and bad women of telenovelas: How to tell them apart using a simple maternity test. *Studies in Latin American Popular Culture, 26*, 97–111.

Viera Cherro, M. (2012). Inequidades múltiples y persistentes en el campo de la reproducción asistida. *Revista de Antropología Social, 21.* https://doi. org/10.5209/rev_raso.2012.v21.40058.

Viera Cherro, M. (2015). Sujetos y cuerpos asistidos Un análisis de la reproducción asistida en el Río de la Plata. *Civitas - Revista de Ciências Sociais, 15*(2), 350. https://doi.org/10.15448/1984-7289.2015.2.17157.

Weihe Edge, B. (2014). Infertility on E! Assisted reproductive technologies and reality television. *Feminist Media Studies, 14*(5), 873–876. https://doi.org/10. 1080/14680777.2014.952875.

Wentzell, E. A. (2013). *Maturing masculinities: Aging, chronic illness, and Viagra in Mexico.* Durham: Duke University Press.

Winner, L. (1980). Do artifacts have politics? *Daedalus, 109*, 121–136.

7

Contemplating a Repronational Portrait

It is a hot midsummer day in Mexico City; I am at the Sonora Metrobus[1] station. The bus arrives and I am pushed in. I struggle to move to one of the ends of the bus. I find a place to stand. I am taller than the women around me, so I can hold on to both handrails like a monkey, one hand on each side. I turn to see the ads displayed on the walls of the bus, on my left an ad for Mary Stopes (an international abortion clinic), and on the right an ad for an AR clinic. I am surprised. I am not surprised to see the Mary Stopes ad, not after the debates and new laws that made abortion a legal and safe option since 2007, I am surprised to see an ad for an AR clinic in the Metrobus! Usually, AR services are framed as available only to the wealthy, and usually the wealthy are not thought of as using public transportation. Soon, however, I come across more ads for AR clinics in spaces where one would not have expected, as, for example, billboards in motorways and in lower-income neighbourhoods. Many of these spaces were where low- to middle-income people lived, so they were not exactly the consumer

---

[1]As of 2018 Line 1 of the Metrobus, where I saw these ads, moves an average of 600,000 people daily.

© The Author(s) 2020
S. P. González-Santos, *A Portrait of Assisted Reproduction in Mexico*,
https://doi.org/10.1007/978-3-030-23041-8_7

of assisted reproduction, but where high-income people transited to go to the airport, to their weekend getaway town, or to the expensive private hospitals. How did AR services get to be a thing worth advertising in these sort of public spaces? In this book, I offered a story of how this happened, of how assisted reproduction was made socially, technically, and legally usable in Mexico. The story I told in this book aimed at highlighting how this process took place within and as part of Mexico's health system development, its political transition, its process of globalisation, and the shifts in its economic regime. In this sense, this story is also a detailed analysis of how Mexico's economy, politics, and culture changed in the course of almost a century. In sum, it is a repronational portrait of Mexico's AR system. In this closing chapter, I want to take one step back to see this portrait from slightly further away.

I began the book by making three analytical moves (see *A National Portrait*). The first was to establish a distinction between the material technologies to assist reproduction, the ARTs (AI, IVF, ICSI, etc.), and the idea of assisting reproduction. This move had the purpose of shifting the attention away from those who were the first to attain a successful birth with ART and instead turn it towards those who were the first to accept the idea of managing and assisting reproduction. This distinction brought to light new actors and ideas which were not usually mentioned in the history of the field, for example, puericulture and esterilología, the physicians who practised these, and the institutions where they were located. The second move was distinguishing between the ARTs and the services (AR services). The purpose of this second move was to be able to see how these technologies were made technically and socially usable, first within the medical community and then within the community of potential users. This distinction also allowed me to highlight the different elements that, as acceptance and usability grew, were added to the services and were described as distinguishing elements between competing clinics (e.g. additional services such as PGD or psychological counselling). The third and final analytical move was viewing the sum of all these elements—ARTs, AR services, AR clinics, specialists, material infrastructure, laws, etc.—as a system. This helped bring together the full range of heterogeneous system builders—and their relations—that have participated in making assisted reproduction socially, technologically, and legally usable in Mexico. The result of these three analytical moves was being able to present a story of how a particular socio-technological system emerged and developed, how a medical speciality professionalised (which is the main topic of the first part of the book, *Origin*), and of the process of commodification of reproduction and infertility (which is the central issue of the second part, *Reproducing Assisted Reproduction*). Finally, it allowed me to identify the style of the system in terms of what it produces and what has powered it.

One of the details of the portrait I focused on was the story of the Mexico's healthcare system, which runs in parallel to the story of the AR system. This fragmented healthcare system has three subsystems: a social security system which is work-based, to which only workers affiliated to specific entities are eligible (IMSS, ISSSTE, PEMEX); a government-run one, to which anybody has access (except those eligible to the work-related system), yet has to be paid for by users; and a private practice which everyone has access to and which users pay as they use either out-of-pocked or through a private insurance company. AR services have been offered in all three branches. (Recently, a new option was introduced, the Seguro Popular, which is a health insurance system that covers basic health issues and is paid by the insured. However, within the period this book covers, this system was not yet fully running.)

The second detail I focused on was the development of puericulture, the medical paradigm that framed reproduction as something medically manageable and no longer as something predetermined and unalterable. I argued that it was puericulture that set the epistemological conditions of possibility for assisted reproduction and ARTs to be later accepted. Puericulture suggested medically managing conception, pregnancy, birth, and the first years of the baby's life through certain behaviours and habits, with the overall purpose of improving the characteristics and health of the baby. In other words, managing reproduction would help produce a healthier individual and thus a healthier population. Puericultura was performed by specially trained gynaecologists and nurses at hygiene centres where pregnant women could go for regular checkups and talks, particularly women with low and middle income. The main proponent of puericulture was the gynaecologist Isidro

Espinosa de los Reyes, who was a teacher at the medical school and an active member of the Mexican Gynaecology and Obstetrics Association. As such, he greatly influenced the physicians who were concerned with infertility and who later founded the Mexican Association of Sterility Studies, the Mexican Association for the Study of Sterility (AMEE).

Origin focused on how the AR system was established by following the story of AMEE, the medical association that today, under a different name (Mexican Association of Reproductive Medicine [AMMR]), gathers the physicians and embryologists who offer AR services. The first noteworthy thing about this association was that, in the beginning, they were openly against artificial insemination, the only ART existing at that moment. How, then, did this association eventually become engaged in high-complexity ARTs? This question guides this first part of the book. The three chapters gathered in *Origin*—*Interest in Sterility*, *Managing Reproduction*, and *Interest in Assisted Reproduction*—traced the development of the association and the medical field, particularly their process of accepting ARTs as part of their field. These chapters look at how this happened alongside the regimes of reproduction operating during the twentieth century. Each of these regimes—puericultura (explored in A National Portrait), esterilología (described in Interest in Sterility), reproductive biology (which is the focus of Chapter 3), and reproductive medicine (detailed in Interest in Assisted Reproduction)—established who could reproduce, how they could reproduce, under which terms, at what cost, and aided by what and by whom. In their own way, they each contributed to the conformation of the AR system.

The purpose of *Interest in Sterility* was to trace how the AMEE, a conservative all-male group of physicians, worked together to build a reputation and a space for esterilología, their emerging speciality. They defined esterilología as the study of sterility and saw it as the medical specialty responsible for treating marital-sterility, which they conceived as a socio-medical condition. In order to trace this process, I analysed the content of their journal, Sterility Studies. Throughout the analysis of two decades of uninterrupted publications, I saw how their conservative views on reproduction, family, and gender set the guidelines to determine who could be an eligible patient—only married heterosexual couples—and which procedures were allowed—only procedures that would not interfere with coitus-based reproduction and were considered to be curing infertility. Their overall objective was to have couples achieve pregnancy, after their intervention, without the need of further medical help. In this chapter, I also argued that, building on puericulture, esterilología contributed to the medicalisation of infertility, framing it as a medically manageable condition and thus, no longer seeing it as a curse, a punishment, or a situation that people were doomed to live with.

AMEE and esterilología transformed during the twentieth century. In *Managing Reproduction*, I described the first stage of this transformation. The chapter was temporarily set at the time when family planning discourse started to be heard in Mexico. Family planning drew attention to contraception, which meant a new way of managing reproduction, not only aiding those who could not conceive but now aiding also those who no longer wished to conceive. At first, AMEE members were concerned with how ideas focused on obstructing reproduction could conflict with their own purposes of aiding reproduction. Then, several of AMEE's members aligned to the family planning programmes, and the association as a whole had to negotiate these tensions. Eventually, these negotiations lead to changes in the epistemic constitution of esterilología and thus emerged a new field, biology of reproduction (biología de la reproducción). In turn, this leads them to change the name of their association in order to reflect this epistemic change. The association was now to be called Mexican Association for the Study of Fertility and Human Reproduction (AMEFRH) and would encompass both ways of managing reproduction: contraception and assisting reproduction. Working simultaneously on both sides of reproduction helped bring funds and attention to reproductive matters, and helped create research programs on reproduction in public institutions such as the National Medical Centre. Hence, the inclusion of both discourses on reproduction allowed the specialty to acquire one more layer of professionalism, granting the AR system more strength.

Chapter 4, *Interest in Assisted Reproduction*, looked at how high-complexity ARTs were made accessible to the medical community interested in fertility issues. For this chapter, I turned to books, conference proceedings, and interviews as a way of tracing this process. This chapter covered the period during which the first AR clinics appeared, the first successful births took place, and the association changed its name to the AMMR. As the name of the association indicates, the epistemic field also changed. What once was esterilología—a field concerned with caring for sterility as defined by the medical profession—then became biology of human reproduction—which focused mostly on managing hormonal levels to either prevent or promote conception—and now was (and still is) reproductive medicine. What distinguishes this new stage is the increased attention given to ARTs over other areas of reproduction such as contraception, menopause, or pregnancy. Evidence of this is provided throughout the chapter, but as an example here I point to two books, one edited in 1999 and the other in 2003, both called Reproductive Medicine. Both of these dedicate a disproportionate number of chapters to assisted reproduction; and the editor indicated that the number of chapters each area has represented the amount of interest and attention being paid to each area (Vázquez-Benítez, 1999, 2003). Hence, assisted reproduction was attracting much attention.

Overall, *Origin* showed how the development of the Mexican AR system was possible because of the dynamic interaction of the areas dedicated to reproduction in the three subsystems of the Mexican healthcare system, the government-run, the social-security one, and the private practice. The style of the Mexican AR system, particularly its usability, has been shaped by the tensions and possibilities resulting from these different branches. The differences between healthcare subsystems mean that people have access to a different combination of possibilities regarding costs and payment schemes, eligibility criteria, and treatment options, and it means that assisted reproduction is placed in a different light in each subsystem. In the social security and the government-run subsystems, assisted reproduction is offered as part of a much wider set of medical interventions, it is offered intermittently, since they need to have funds to be able to offer it, and the patient needs to go through other services prior to accessing the AR service. This blurs the visibility of these subsystems' AR services and its users. However, because these subsystems are very important in as educational sites, its presence in these subsystems has contributed to disseminate knowledge about assisted reproduction among a wide sector of medical students and thus prepare future practitioners in the field. This contrasts with the way assisted reproduction is offered in private practice, where it has become highly visible. Private AR clinics have framed themselves as independent and highly specialised healthcare facilities which, as I showed in the second part of the book, engaged in marketing strategies that helped shed light on them thus adding visibility to their services and depicting a very specific sort of user: the wealthy. This combination of visibilities, with the backdrop of puericulture and the family planning campaigns, results in an overall popular view that infertility and assisted reproduction are only important to and used by the wealthy, while the other segments of the population are commonly thought of as extremely fertile and in need of contraception. All this contributed to making assisted reproduction usable, but this usability is very different for different people.

Origin, left some unanswered which require further attention. For example, more research is required to understand the role played by two more groups of people interested in biology of reproduction. One was the group of researchers located at the Instituto Nacional de Nutrición Salvador Zubirán (INNSZ) and the other was the group of people organised in the professional association Academia de Investigadores en Biología de la Reproducción (AIBIR). These two groups hardly came up in my research, maybe due to the focus of this book, which was on how the professional field of AR developed by following the life of the AMMR. Nonetheless it is important to ask and find out what role they played in the overall AR system. Another under-research aspect concerns the role of women. As I have mentioned several times throughout the book, during the period that this book covers, roughly from 1930 to 2010, very few Mexican women were mentioned, and when they were, these were only "one-time" mentions (two published a paper in Sterility Studies, and one participated in the creation of RedLARA's registry). Why was this the case? Where were all the women?

The Part II of the book, *Reproducing Assisted Reproduction*, looked at how the AR system disseminated the idea, among both practitioners and within a large population of potential users, that assisted reproduction was an acceptable way of managing infertility and of forming a family. Analytically, I framed assisted reproduction and ARTs as cultural novelties. In brief, cultural novelties are new ways of doing-naming-understanding, and as such they need to be accepted in order for them to be incorporated into culture. This also means that potential users (in this case practitioners and potential patients) have to know about it, have a need for it, have it available and accessible in some shape or form, and need to be able to articulate it with their existing culture. I traced this process throughout the two chapters that make up this section: *The Universe Is Expanding* and *The Discursive Landscape*.

In *The Universe Is Expanding*, I looked at how knowledge and experience with ARTs were communicated and accepted among a larger sector of gynaecologists. This acceptance meant that there were more books being published in the field, more students seeking to specialise in it, and eventually, more clinics offering these services. While in the late eighties, there were only three AR services, by the end of the century there were already seventeen, and by 2018 there were, according to some estimates, 100 clinics across 21 states (Cofepris, 2019). With the exception of a handful, these were all private clinics. Soon, all these clinics came to look alike, establishing a particular category which I called the AR clinic. This particular type of clinic is distinguishable in its technological and spatial infrastructure. It has, for example, an embryo and gametes lab adjacent to the operating room, a sperm collecting room, and very specific environmental conditions. Its staff composition is also unique; it needs, at least, nurses, administrative personnel, embryologists, and gynaecologists, most of them working as a team where all overlook the cycles of all patients. These clinics are also different from other healthcare practitioners' offices in that they employ marketing strategies to promote their services. Due to all this, establishing and running an AR clinic requires many things: a special knowledge set, a trained staff, large sums of money, high-tech equipment, and a minimum and maximum of cycles that need to be carried out and organised in very specific ways. In this chapter, I also traced how the field consolidated its epistemic and normative make-up, for example, by obtaining official recognition as a subspecialty, by establishing alliances, and participating in the Latin American AR Network (RedLARA).

In Chapter 6, *The Discursive Landscape*, I looked at how assisted reproduction was disseminated among potential users through a discursive landscape confirmed by different media spaces (press, television, radio), public spaces (advertisements in public transport), and semi-public spaces (trade shows and information talks). Overall, assisted reproduction was framed, throughout these spaces, as comprised of a series of a para-natural technologies capable of helping to produce marianistic mothers and heteronormative families. Paranatural technologies are those which are famed as imitating nature while simultaneously going beyond it to improve where it fails. This framing was particularly used by physicians who were explaining the procedures to potential patients at information sessions, during their talks at the trade show, and in interviews or comments given to the media. This framing made assisted reproduction socially usable because it de-emphasised the artificiality and riskiness, while it emphasised the technological and controlled aspect of these procedures, and thus made them seem natural, safe, and successful. Assisted reproduction can be seen as participating in the (re)production of people, children, families, fathers, mothers, citizens, donors, siblings, etc., but in the discursive landscape analysed in this book, these paranatural technologies were producing mainly heteronormative marianistic mothers.

Given the importance of motherhood in Mexican culture, I argue that this framing has facilitated its acceptance. Assisted reproduction could have been described as producing children or larger families (as described, for example, by Almeling 2011). If so, it would have conflicted with the embedded idea that Mexico needs to reduce its population size (even when family planning campaigns were nearly absent between 2000 and 2012, due to the conservative governments of Fox and Calderón). Likewise, if it were framed as producing fathers, the discourse would have had to focus on men as infertile beings, which in turn could be associated with erectile dysfunction, and thus could have conflicted with macho culture (cf. Wentzell, 2013). But because it is marianistic mothers what the Mexican AR system is producing, assisted reproduction acquired force and was deemed acceptable. Marianistic motherhood is the cultural understanding of motherhood in Mexico. This understanding has, at its core, the idea that a good mother sacrifices everything for her family—children and husband. AR users were women who did everything to become mothers; therefore, their engagement with these technologies and the sacrifices that they imply grants them the status of marianistic mothers.

The last aspect analysed in Chapter 6 was the commercial touch. The AR system has expanded mostly as a private practice with clinics competing for a limited sector of users—limited because of economic aspects, not due to lack of need nor rejection. This has resulted in the emergence and proliferation of publicity campaigns. The discursive landscape is full of messages concerning costs, discounts, and payment schemes alongside messages of guarantees, hope, and promises that are difficult to keep. These messages are loud against a silent backdrop, where risk and other forms of kinship (e.g. adoption) can hardly be heard. This has also been possible due to a lack of regulation in terms of how and what can be advertised, a particularly difficult matter after the emergence of the Internet and of social media (Facebook, Instagram, Twitter, and the likes). This commercial touch has resulted in the commodification of infertility and reproduction, and the marketisation of assisted reproduction. As commented in the chapter, some people thought that AR services have reacted to competition as other commodities do, making them less expensive. This needs further study to see if prices have actually gone down or at least have not increased much and if this is related to competition or to other phenomena.

This second part also left questions unanswered which need further research. First, in this book I delved into how this particular medical specialty understood infertility, the assigned role of the female and the male bodies in the process of reproduction, and thus their ways of studying and reporting the data concerning infertility and the success rates of ARTs. However, more analysis in this area is needed, for example, to understand why, within the field of AR, women's bodies, histories, and lives are seldom taken into account. With the exception of their chronological age, the rest of their medical body (other health diagnosis related to hormones and reproduction) is not considered. The same question is applicable to men: why are their bodies, health histories, lifestyles, etc. not considered? Between the 1950s and the 1980s, some attention was given to men as reproductive beings, first as part of the questions addressed by esterilología and then mostly within the family planning campaigns. But by the first decade of the twenty-first century this attention had diminished to the point that in the RedLARA's registry, men are not even considered (see Sect. "Documenting as Narrative" in Chapter 5).

Second, more attention needs to be given to the embryologist. Within this profession, there is a more balanced gender representation, a more collaborative practice, and it is in their area of expertise where much of the recent advances are being done. Hence, better understanding of how this profession has developed, specialised, and professionalised would be important. Third, given the costs and administrative practices needed to establish and maintain an AR clinic, it would be important to understand how these elements have participated in shaping how AR services are offered, and how the specialty is practised. Understanding how the economic aspect of AR services shapes practice could help to better understand how assisted reproduction is commodifying reproduction (cf. Almeling, 2011). Fourth, as illustrated throughout the book, the AR system has been built and developed through the interaction of healthcare professionals, embryologists, nurses, users, etc., both nationally and internationally. However, these have not been the only system builders. Lawyers, marketing experts and content developers, equipment and supplies distributors, and administrators have been equally important. More work needs to be done on the emergence, consolidation, and role of these actors within the AR system. Fifth, the interaction between Mexico and Spain, on the one hand, and Mexico and Latin America, on the other, has also been relevant to this process. Further research into the characteristics and outcomes of these interactions is needed. Likewise, given the role played by RedLARA in Mexico and the rest of Latin America, it would be interesting to work on a more detailed understanding of how this network operates, why they do so in the way they do, and what information they obtain and produce (see Chapter 5). Overall, more research in these areas would also help to better understand the AR global market. In recent years, there has been some attention given to Mexico and its relation to the assisted reproduction global market, but only concerning the use of surrogacy by foreign couples and individuals (Olavarría Patiño, 2018; Schurr, 2017; Walmsley & Schurr, 2014). Articulating these very focused studies with the few that have a broader look of the field of assisted reproduction, as well as with those focusing on the legal and bioethical aspects, would be useful. In other words, more collaborative and interdisciplinary work to bring the different areas of research on assisted reproduction in México would be welcomed.

This book has suggested how assisted reproduction has been made socially and technically usable and how, by using it, ideas and practices concerning kinship, reproductive health and illness, gender roles, and the relationship between science and religion have been questioned, explored, reproduced, and reconfigured. As such, it has sought to contribute to the scholarship on commodification that considers this as a process that can "proceed in many different ways and be imbued with many different meanings" (Almeling, 2011: 5). The processes this book looks at are very much ongoing and many aspects are still in need of research; hence, this is not a full stop, just a pause. Hopefully more research is on its way.

The final point made in this book that I would like to highlight is the speculative reflection with which I concluded Chapter 6: Could it be that we are witnessing the gestation of two approaches of assisted reproduction? One restorative—which aims at faithfully reproducing the heteronormative family—and one re-configurative—which has the power of subverting this family structure. The difference between these two approaches does not rest solely on the family they produce, but also in the normative, epistemic, and maybe even material make-up of each one. The specialised knowledge needed by a practitioner who is caring for a heterosexual couple with a diagnosis of infertility due to low sperm count and endometriosis is very different from that of a practitioner caring for, say, a same-sex couple where one might be transgender. In other words, the different kinds of family structure that assisted reproduction opens up for does not just stop there; assisted reproduction, as a platform technology (Franklin, 2013), is also opening up different professional possibilities, one might be the birth of a new medical approach to this particular subspecialty.

I am awestruck by how the creation of a new human being now implies an assemblage of heterogeneous elements that have painted this repronational portrait: expensive equipment and supplies that have been mobilised from a variety of cities and countries; multimedia marketing and publicity schemes that deploy discounts, raffles, infomercials, singing sperm, highway billboards, sperm-shaped USBs and pens, and financing schemes no different from those offered to buy a house and available also for cosmetic plastic surgery; messages suggesting that controlling reproduction is possible, offering a 100% guarantee, building hopes that people hang on to while their bodies, savings, and marriages cripple. I cannot stop being surprised, amazed, and a bit concerned when I see how the process of reproduction becomes more mediated, more commercial, more commodified, and the conditions for the possibility for more intervention, control, design, and planning increase. As recent events have suggested, there is also the possibility for new kinds of accidents and surprises, for example, the recent nitrogen drop in the cryopreservation tanks at two clinics in the USA, where thousands of embryos and gametes were kept and thus lost (Gray, 2018), the birth of the first babies conceived after mitochondrial replacement therapy in Mexico (González Santos, Stephens, & Dimond, 2018; Zhang et al., 2016), and then the birth of twins who, allegedly, were subject to gene therapy using CRISPR (Arey, 2018).

I want to highlight the need for a reflexive way of incorporating assisted reproduction into our lives, both materially and symbolically; the need to take care of, and responsibility for, all actors involved, particularly the future generations and society as a whole; the need for remembering that there are and have always been multiple ways of forming families, that birthing a baby is not the only way to make kin (invoking Haraway's claim to make kin, not babies); the need to never forget that artefacts (and assisted reproduction is an artefact and a system of artefacts) have politics and that our actions make world. This book began, long before it started, with the question concerning what makes us human. Today, after two decades of studying the cyborgification of our bodies and societies, this book closes with the question of what makes our collective, a human collective and if this can ever be lost.

Detail of a National Portrait

I find myself sitting still, on a high chair, against the wall. The lights were slightly dimmed. In front of me, two bodies dressed in scrubs: heads covered, mouths covered, bodies covered, shoes covered; all I could see were their eyes behind their glasses. They were preparing two embryo transfers. One was thawing an embryo, while the other was preparing the tools, opening packages, taking out things, and setting them in the right order and in the right place. A long loud artificial hum accompanied the radio morning news and these embryologists' silence. The presenter's voice coming out of the speaker was talking about traffic jams, the earthquake's casualties, a bloody assassination, the blocking of streets, Trump's policies, Peña Nieto's corruption; these words of crisis, conflict, despair, anger, and turmoil enveloped the room. The two embryologists where working, concentrated, slowly moving from microscope to incubator, from incubator to bench, from bench to hatch, and back again. They held a respectful and caring silence; such loud silence in the middle of all the noise. (In my head: "All the while the world is turning to noise".)

Then the voice of one of the embryologist: "*Voy*" (coming).

She places the catheter straight up, taking both ends with her hands and carefully holding it close to her chest, she moves towards the door. The silence intensifies, I hold my breath, and the news presenter keeps on recounting death. In perfect coordination, at the precise moment, the other embryologist opens the door. In the hallway, you see the door to the operating room open. She crosses the doorway into the operating room. The door is shut. We wait. Advertisements.

She comes back in. The door is shut again. She goes to the bench and looks into the microscope. The silence is heightened with a "*todo bien*" (all OK).

(In my head, a voice asks: Is it? Is everything really ok now?)

References

Almeling, R. (2011). *Sex cells: The medical market for eggs and sperm*. Berkeley: University of California Press.

Arey, W. (2018, December 31). Web roundup: CRISPR babies and bioethics. *Somatosphere*. Retrieved January 12, 2019, from http://somatosphere.net/2018/12/web-roundup-crispr-babies-and-bioethics.html.

Cofepris (Comisión Federal para la Prevención del Riesgo Sanitario). (2019). *Listado de establecimientos Autorizados para reproducción Asistida*. Accessed 22 July 2019. https://www.gob.mx/cms/uploads/attachment/file/439319/SEASS_RA.pdf.

Franklin, S. (2013). *Biological relatives: IVF, stem cells, and the future of kinship.* Durham: Duke University Press.

Gray, M. (2018, March 28). *More than 4,000 eggs and embryos lost in Cleveland fertility clinic tank failure.* Retrieved February 15, 2019, from CNN website: https://edition.cnn.com/2018/03/27/health/cleveland-fertility-clinic-eggs-embryos-lost/index.html.

González Santos, S. P., Stephens, N., & Dimond, R. (2018). Narrating the first "three-parent baby": The initial press reactions from the United Kingdom, the United States, and Mexico. *Science Communication, 40*(4), 419–441. https://doi.org/10.1177/1075547018772312.

Olavarría Patiño, M. E. (2018). La gestante sustituta en México y la noción de trabajo reproductivo. *Revista Interdisciplinaria de Estudios de Género de El Colegio de México, 4,* 1–31.

Schurr, C. (2017). From biopolitics to bioeconomies: The art of (re-)producing white futures in Mexico's surrogacy market. *Environment and Planning D: Society and Space, 35*(2), 241–262. https://doi.org/10.1177/0263775816638851.

Vázquez-Benítez, E. (Ed.). (1999). *Medicina Reproductiva en México* (Primera). México, D.F., Mexico: Manual Moderno JGH Editores S.A. de C.v.

Vázquez-Benítez, E. (Ed.). (2003). *Medicina Reproductiva* (2º). México, D.F., México: Editorial Manual Moderno, AMMR.

Walmsley, H., & Schurr, C. (2014). Reproductive tourism booms on Mexico's Mayan Riviera. *International Medical Travel Journal, 3.*

Wentzell, E. A. (2013). *Maturing masculinities: Aging, chronic illness, and Viagra in Mexico.* Durham: Duke University Press.

Zhang, J., Liu, H., Luo, S., Chavez-Badiola, A., Liu, Z., Yang, M., … Huang, T. (2016). First live birth using human oocytes reconstituted by spindle nuclear transfer for mitochondrial DNA mutation causing Leigh syndrome. *Fertility and Sterility, 106*(3), e375–e376. https://doi.org/10.1016/j.fertnstert.2016.08.004.

Appendix

Technologies to Assist Reproduction

Assisted Hatching

Before an embryo can attach to the wall of the womb, it has to break out or "hatch" from its outer layer. Therefore, in some cases, prior to being transferred back to the womb, a hole is made in the embryo's zona pellucida (the outer layer of the embryo) or it is thinned using acid, laser or mechanical methods to help it hatch and implant.

Gamete Intra-Fallopian Transfer (GIFT)

After selecting the healthiest eggs and sperm, they are placed together in the woman's fallopian tubes. Fertilisation therefore takes place within the body.

S. P. González-Santos, *A Portrait of Assisted Reproduction in Mexico*,
https://doi.org/10.1007/978-3-030-23041-8

Intra-Cytoplasmic Sperm Injection (ICSI)

The procedure for ICSI is similar to that for IVF, but instead of fertilisation taking place in a dish and with little assistance, in ICSI the embryologist selects one single sperm from the sample and injects it directly into the ova using a potent microscope. Then the injected ova are left to rest, if fertilisation occurs the rest of the procedures takes place like in IVF.

Intrauterine Insemination (IUI)

Using different techniques, the fast-moving sperm are separated from slower or non-moving sperm. Then a concentration of capacitated and washed fast-moving sperm is placed into the woman's womb close to the time of ovulation, around the middle of the ovulation cycle, when the ova is released from the ovary.

In Vitro Fertilisation (IVF)

Usually, the IVF process begins with ovarian stimulation to obtain a larger number of ova than with a non-assisted cycle. The number of days women take the drugs depends on the type of drug cocktail used. Throughout the drug treatment, the doctor monitors the ova development progress using vaginal ultrasound scans and, possibly, blood tests. The purpose of these drugs is to suppress the natural ovulation cycle in order to control it and to increase the number of ova produced. Drugs are given to promote the maturation of follicles, then, once the follicles reach the desired size, and 34–38 hours before they are due to be collected, a hormone injection is given to help them finish their maturing process. Once they are collected from the ovaries, usually by ultrasound guidance and under sedation, they are placed in a Petri dish with a high concentration of capacitated sperm. Sperm and ova are cultured there for between 16 and 20 hours; then they are checked to see if fertilisation occurred. If fertilisation occurs, the resulting embryo is left to mature for a couple of days longer before being transferred to the woman's womb. If possible, the embryos considered to have the best morphological qualities are chosen for transferral. In some cases, another cocktail of drugs is given to prepare the lining of the womb for embryo transfer, to help during the implantation process and to help throughout pregnancy.

In Vitro Maturation (IVM)

Ova are collected from the ovaries when they are still immature and left to mature in the laboratory before being fertilised. With this technique, compared to conventional IVF, women do not need to take as many drugs before ova can be collected.

Sims-Hühner Post-coital Test

This procedure was developed by Sims (in 1868) and by Hühner (in 1912), who first used it to examine infertile couples. This test is used to evaluate the fertility of both female (the quality of the mucus) and male (the quality and number of sperm). The objective of the test is to examine the cervical mucus to asses its functionality, and to see the quality of the sperm that penetrated. The test consists of sampling the cervical mucus between 2 and 8 hours after intercourse. The sample is taken using a Papanicolaou pipette and analysed. First, they see the aspect, viscosity, and elasticity (spinnbarkeit) of the mucus, and then they estimate the amount of sperm in the sample and evaluate their condition (motility) (Giner, Merino, Luna, & Aznar, 1974). The rationale behind this test is that, the secretions of the cervix guide, nourish, and protects sperm, so a good quality mucus will favour sperm survival and arrival to the egg for fertilisation (Moghissi 1976). This test is also known as espermatobioscopia indirecta, while the espermatobioscopia directa (EBD) is done on sperm collected after masturbation.

Sperm Collection and Manipulation Techniques

Sperm can be obtained by masturbation, by vaginal intercourse with the use of specially designed condoms, or surgically. There are various options for surgically obtaining sperm, these are: microsurgical epididymal sperm aspiration (MESA), testicular sperm extraction (TESE), or puncturing the seminal vesicle or the vas deferents. Once fresh sperm has been obtained, it is washed and spun at a high speed in order to select the most active sperm. The sample can then be frozen for future use or used immediately for insemination of fertilisation.

Glossary

AMGO Asociación Mexicana de Ginecología y Obstetricia [Mexican Association of Gyneacology and Obstetrics]

AMI Asociación Mexicana de Infertilidad [Mexican Infertility Association]

AMMR Asociación Mexicana de Medicina de la Reproducción [Mexican Association of Reproductive Medicine]

CCF Código Civil Federal [Federal Civil Code]

CENATRA Centro Nacional de Transplantes [National Transplant Centre]

CNEGySR Centro Nacional de Equidad de Género y Salud Reproductiva [National Gender Equity and Reproductive health Centre]

COFEPRIS Comisión Fedral para la Protección Contra Riesgos Sanitarios [Federal Commission Against Sanitary Risk]

COMEGO Consejo Mexicano de Ginecología y Ostetricia [Mexican Council of Gynaecologists and Obstetrics]

CONAMED Consejo Nacional de Arbritaje Médico [National Council for Medical Arbitration]

CONAPO Consejo Nacional de Población [National Population Council]

CPF Código Penal Federal [Federal Criminal Code]

IMSS Instituto Mexicano del Seguro Social [Mexican Social Security Institute]

© The Editor(s) (if applicable) and The Author(s) 2020
S. P. González-Santos, *A Portrait of Assisted Reproduction in Mexico*,
https://doi.org/10.1007/978-3-030-23041-8

INMEGEN Instituto Nacional de Medicina Genómica [National Institute of Genomic Medicine]

INNSZ Instituto Nacional de la Nutrición Salvador Zubirán [National Institute of Nutrition]

ISSSTE Instituto de Seguridad y Servicios Sociales de los Trabajadores del Estado [Social Security Institute for the State Workers]

LGP Ley General de Población [General Population Law]

LGS Ley General de Salud [General Health Law]

PEMEX Petróleos Mexicanos [National Petrol Company]

Red LARA Red Latino Americana de Reproducción Asistida [Latin American Assisted Reproduction Network]

SEDENA Secretaría de la Defensa Nacional [Ministry of Defence]

SS Secretaría de Salud [Ministry of Health]

UAM Universidad Autónoma Metropolitana [Metropolitan Autonomous University]

UNAM Universidad Nacional Autónoma de Mexico [National Autonomous University of Mexico]

References

Abalovich, M., Mitelberg, L., Allami, C., Gutierrez, S., Alcaraz, G., Otero, P., & Levalle, O. (2007). Subclinical hypothyroidism and thyroid autoimmunity in women with infertility. *Gynecological Endocrinology, 23*(5), 279–283. https://doi.org/10.1080/09513590701259542.

Abusief, M., Hornstein, M., & Jain, T. (2007). Assessment of United States fertility clinic websites according to the American Society for Reproductive Medicine (ASRM)/Society for Assisted Reproductive Technology (SART) guidelines. *Fertility and Sterility, 87*(1), 88–92. https://doi.org/10.1016/j.fertnstert.2006.05.073.

Acevedo, E. (2003, Marzo). Alfonos Gutiérrez Najara: Ayudando a la vida a dar vida. *Medicos de México.* Año 1. No. 10. 12.

Adame Goddard, J. (2004). El régimen revolucionario del matrimonio civil. In *El matrimonio civil en MÉXICO (1859–2000)* (pp. 35–82). Instituto de Investigaciones Jurídicas - UNAM. Retrieved from https://archivos.juridicas.unam.mx/www/bjv/libros/3/1362/4.pdf.

Agostini, C. (2007). Las mensajeras de la salud. Enfermeras visitadoras en la Ciudad de MÉXICO durante la década de los 1920. *Estudios de Historia Moderna y Contemporánea de México, 33.*

Ahedo, A. (2017, December 12). Más de 7 millones de personas visitan la Basílica de Guadalupe. *El Universal.* Retrieved from https://www.eluniversal.com.mx/metropoli/cdmx/mas-de-7-millones-de-personas-visitan-la-basilica-de-guadalupe.

© The Editor(s) (if applicable) and The Author(s) 2020
S. P. González-Santos, *A Portrait of Assisted Reproduction in Mexico,*
https://doi.org/10.1007/978-3-030-23041-8

267

Ahued Ahued, R. J. (2004). Semblanza del Doctor ISIDRO ESPINOSA DE LOS REYES. *Perinatología y Reproducción Humana, 18*(4), 205–207.

Alanís, M. (2015). MÁS que curar, prevenir: surgimiento y primera etapa de los Centros de Higiene Infantil en la Ciudad de MÉXICO, 1922–1932. *História, Ciências, Saúde-Manguinhos, 22*(2), 391–409. https://doi.org/10.1590/S0104-59702015005000004.

Alba-Hernandez, F. (Ed.). (1976). *La población de MÉXICO.* El Colegio de México, México: Centro de Estudios Económicos y Demográficos, El Colegio de México.

Albury, R. (1999). *The politics of reproduction: Beyond the slogans.* Crows Nest, NSW: Allen & Unwin.

Allende, A. C. (2012). Telenovelas en México. Nuestras íntimas extrañas. *Comunicación y Sociedad, Nueva época* (18), 205–210.

Almeling, R. (2011). *Sex cells: The medical market for eggs and sperm.* Berkeley: University of California Press.

Alvarado Durán. A. (1999). Ciencia y Bioética. Cap. 73 (p. 387). In E. Vázquez Benítez (Ed.), *Medicina Reproductiva en México* (Primera). Mexico: JGH Editores S.A. de C.V.

Alvarado Durán, A. (2003a). Bioética en Reproducción Asistida. Sección XVI Bioética en reproducción. Cap. 69 (p. 521). In E. Vázquez Benítez (Ed.), *Medicina Reproductiva* (2º). México, D.F., Mexico: Editorial Manual Moderno, AMMR.

Alvarado Durán, A. (2003b). Riesgos legales en los estudios y tratamientos de esterilidad. Sección XVII Aspectos Legales de los estudios y tratamientos relacionados con la reproducción. Cap. 71 (p. 535). In E. Vázquez Benítez (Ed.), *Medicina Reproductiva* (2º). México, D.F., Mexico: Editorial Manual Moderno, AMMR.

Álvarez Bravo, A. (1952). Plan de Organización de la lucha contra la esterilidad conyugal. *Estudios sobre Esterilidad, 3*(1), 1.

Alvarez Lezama, F. J. (1964, Mayo–Ago). El incremento demográfico y la planeación familiar. *Estudios Sobre Esterilidad, XV*(2), 64.

Álvarez Navarro, M. (1999). La Bioética en Reproducción. In E. Vázquez Benítez (Ed.), *Medicina Reproductiva en México* (Primera). Mexico: JGH Editores S.A. de C.V.

Álvarez Navarro, M. (2003). Introducción. Sección XVI Bioética en reproducción (p. 511) In E. Vázquez Benítez (Ed.), *Medicina Reproductiva* (2º). México, D.F., México: Editorial Manual Moderno, AMMR.

Appadurai, A. (1990). Disjuncture and difference in the global cultural economy. In M. Featherstone (Ed.), *Global culture: Nationalism, globalization and modernity* (pp. 295–310). London: Sage.

Arey, W. (2018, December 31). Web roundup: CRISPR babies and bioethics. *Somatosphere*. Retrieved January 12, 2019, from http://somatosphere.net/2018/12/web-roundup-crispr-babies-and-bioethics.html.

Arranz-Lara, L. (2001). El deseo de maternidad en pacientes sujetas a tratamientos de reproducción asistida: una propuesta de psicoterapia. *Perinatol Reprod Hum, 15*(2), 6.

Arredondo-Rivera, R. M., & Juárez-Sánchez, J. M. (2009, Agosto). Aporte histórico y revolucionario de un egresado de la FQ. *Gaceta Facultad de Química*, VII época, No. 54, México: UNAM.

Arteaga Elizondo, O. (1961, Mayo–Agosto). Consideraciones Psicológicas, morales y jurídicas sobre la inseminación artificial humana. *Estudios Sobre Esterilidad, 7*(2), 59.

Arzac, P. (1956, Enero–Abril). Infrome del Presidente de la Asociación Mexicana de Estudios de la Esterilidad correspondiente al periodo 1953–1955. *Estudios sobre Esterilidad, VII*(1), 49.

Arzac, J. P. (1959). Meditaciones sobre la Asociación Mexicana para Estudios de la Esterilidad. *Estudios Sobre Esterilidad, 10*(2), 147.

Asakura, H. (2005). Cambios en significados de la maternidad: la emergencia de nuevas identidades femeninas. Un estudio de caso: mujeres profesionistas de clase media en la Ciudad de México. In M. Torres (Ed.), *Nuevas maternidades y derechos reproductivos* (pp. 33–59). México: El Colegio de México.

Asch, R. H., Balmaceda, J. P., Ellsworth, L. R., & Wong, P. C. (1986). Preliminary experiences with gamete intrafallopian transfer (GIFT)**Recipient of the 1985 Squibb Award. Presented at the Forty-First Annual Meeting of The American Fertility Society, September 28 to October 2, 1985, Chicago, IL. *Fertility and Sterility, 45*(3), 366–371. https://doi.org/10.1016/s0015-0282(16)49218-6.

Austin, W. (2001). Using the human rights paradigm in health etics: The problems and the possibilities. *Nurse Ethics, 8*(3), 183–195.

Baker, J. H. (2012). *Margaret Sanger: A life of passion* (Reprint ed.). New York: Hill and Wang.

Barbey, C. (2017). Evidence of biased advertising in the case of social egg freezing. *The New Bioethics, 23*(3), 195–209. https://doi.org/10.1080/20502877.2017.1396033.

Basten, S. (2009). *Mass media and reproductive behaviour: Serial narratives, soap operas and telenovelas.* The future of human reproduction. Retrieved from http://citeseerx.ist.psu.edu/viewdoc/download?doi=10.1.1.701.5228& rep=rep1&type=pdf.

Basten, S. (2010). Television and fertility. *Finnish Yearbook of PopulationResearch, XLV,* 67–82.

Beard, L. J. (2003). Whose life in the mirror?: Examining three Mexican telenovelas as cultural and commercial products. *Studies in Latin American Popular Culture, 22,* 73–88.

Becker, G., Butler, A., & Nachtigall, R. D. (2005). Redemblance talk: A challenge from parents whose children were conceived with donor gametes in the US. *Social Science and Medicine, 61,* 1300–1309.

Bharadwaj, A. (2000). How some Indian baby makers are made: Media narratives and assisted conception in India. *Anthropology & Medicine, 7*(1), 63–78.

Bharadwaj, A. (2006a). Sacred conceptions: Clinical theodicies, uncertain science, and technologies of procreation in India. *Culture, Medicine and Psychiatry, 30*(4), 451–465. https://doi.org/10.1007/s11013-006-9032-0.

Bharadwaj, A. (2006b). Sacred modernity: Religion, infertility, and technoscientific conception around the globe. *Culture, Medicine and Psychiatry, 30*(4), 423–425. https://doi.org/10.1007/s11013-006-9030-2.

Bharadwaj, A. (2016). *Conceptions: Infertility and procreative modernity in India.* New York: Berghahn Books.

Bijker, W. E., Hughes, T. P., & Pinch, T. (2012). *The social construction of technological systems: New directions in the sociology and history of technology.* Cambridge, MA: MIT Press.

Borrero, C., Ord, T., Balmaceda, J. P., Rojas, F. J., & Asch, R. H. (1988). The gift experience: An evaluation of the outcome of 115 cases. *Human Reproduction, 3*(2), 227–230. https://doi.org/10.1093/oxfordjournals.humrep.a136682.

Caldwell, J. C. (2001). The globalization of fertility behavior. *Population and Development Review, 27,* 93–115.

Çalışkan, K., & Callon, M. (2009). Economication, part 1: Shifting attention from the economy towards processes of economization. *Economy and Society, 38*(3), 369–398. https://doi.org/10.1080/03085140903020580.

Callon, M. (1987). Society in the making: The study of technology as a tool for sociological analysis. In W. E. Bijker, T. P. Hughes, & T. Pinch (Eds.),

The social construction of technological systems: New directions in the sociology and history of technology. Cambridge, MA: MIT Press. Retrieved from http://search.ebscohost.com/login.aspx?direct=true&scope=site&db=nlebk&db=nlabk&AN=457446.

Callon, M., Law, J., & Rip, A. (Eds.). (1986). *Mapping the dynamics of science and technology.* London: Palgrave Macmillan. https://doi.org/10.1007/978-1-349-07408-2.

Canto de Cetina, T. E. (1994). Programas de biología de la reproducción en México y particularmente en Yucatán. *Revista Biomed, 5*(3), 141–150.

Cardozo, E., Pavone, M. E., & Hirshfeld-Cytron, J. E. (2011). Metabolic syndrome and oocyte quality. *Trends in Endocrinology & Metabolism, 22*(3), 103–109. https://doi.org/10.1016/j.tem.2010.12.002.

Carreño-Meléndez, J., Morales-Carmona, F., Sánchez-Bravo, C., González-Campillo, G., & Martínez-Ramírez, S. (2003). En parejas estériles por factor masculino y femenino. *Perinatol Reprod Hum, 17*(2), 11.

Castelazo Ayala, L. (1959, Mayo–Agosto). Discurso del Dr. Castelazo Ayala. Presidente entrante de la AMEE. *Estudios Sobre Esterilidad, 10*(2), 5.

Castelazo Ayala, L. (1960). Importancia de la instrucción prenupcial. *Estudios sobre Esterilidad, 11*(3), 142.

Castro, E. (1959, Mayo–Agosto). Un urólogo en deuda con la Asociación Mexicana para Estudios sobre la Esterilidad. *Estudios Sobre Esterilidad, 10*(2).

Cerón, R. (2007). La ciencia vence a la infertilidad. *El Universal,* March 1. Section: Salud.

Challoner, J. (1999). *The baby makers: The history of artificial conception.* London: Channel 4 Books.

Chavarría Olarte, M. A., Gallegos Cigarroa, A., & Reyes Fuentes, A. (2018). Unidad de Investigación Médica en Medicina Reproductiva. In T. Miguel Ortega, J. de Jesús Arriaga Dávila, A. C. Sepúlveda Vildósola, F. A. Salamanca Gómez, & C. Martínez Castuera Gómez (Eds.), *Contribuciones del IMSS a la medicina mundial. Pasado, presente y futuro. Instituto Mexicano del Seguro Social.* México: IPN. Retrieved from https://www.ipn.mx/assets/files/secgeneral/docs/SG/Libro%20IMSS%2075%20a%C3%B1os.pdf.

Chávez-Courtois, M. L. (2004). Presencia de esterilidad: Actores o sujetos en la actualidad. *Escuela Nacional de Antropología e Historia (ENAH). Cuicuilco, 11*(31), 1–17.

Chávez-Courtois, M. L. (2014). Presencia de esterilidad: actores o sujetos en la actualidad. *Cuicuilco. Revista de Ciencias Antropológicas, 11*(31), 191–206.

Chavez Villasana, A. (1964, Mayo–Ago). El incremento demográfico y la planeación familiar. *Estudios Sobre Esterilidad, XV*(2), 72.

Chen, S. H., Pascale, C., Jackson, M., Szvetecz, M. A., & Cohen, J. (2016). A limited survey-based uncontrolled follow-up study of children born after ooplasmic transplantation in a single centre. *Reproductive BioMedicine Online, 33*(6), 737–744. https://doi.org/10.1016/j.rbmo.2016.10.003.

Cícero Sabido, R., Padua Gabriell, A., Rodriguez Martinez, H., Toledo, B., & Yáñez Villar, A. (1986). Efectos del terremoto del 19 de septiembre de 1985 en el Hospital General de la Ciudad de México. Algunas consideraciones. *Salud Públia México, 28*, 521–526.

Cofepris (Comisión Federal para la Prevención del Riesgo Sanitario). (2019). *Listado de establecimientos Autorizados para reproducción Asistida.* Accessed on 22 July 2019. https://www.gob.mx/cms/uploads/attachment/file/439319/SEASS_RA.pdf.

Cohen, J., & Alikani, M. (2013). Evidence-based medicine and its application in clinical preimplantation embryology. *Reproductive BioMedicine Online, 27*(5), 547–561. https://doi.org/10.1016/j.rbmo.2013.08.003.

Cohen, J., Scott, R., Schimmel, T., Levron, J., & Willadsen, S. (1997). Birth of infant after transfer of anucleate donor oocyte cytoplasm into recipient eggs. *The Lancet, 350*(9072), 186–187. https://doi.org/10.1016/S0140-6736(05)62353-7.

Collier, S. J., & Lakoff, A. (2004). Ethics and the anthropology of modern reason. *Anthropological Theory, 4*(4), 419–434. https://doi.org/10.1177/1463499604047919.

Colmeiro-Lafortet, C. (1952). La falta de orgasmo como causa de esterilidad. *Estudios sobre Esterilidad, III*(2) (Abril de 1951), 57.

Conde Manjarerez, B. (1967, Septiembre–diciembre). Utilidad del método citológico en la amenorrea. *Estudios Sobre Esterilidad, XVIII*(3), 118.

Consejo Nacional de Población. (1999). Introducción. Veinticinco años de planificación familiar en México. *La situación demográfica de México.* México: Consejo Nacional de Población.

Cook, R. J., Dickens, B. M., & Fathalla, M. F. (2003). *Reproductive health and human rights.* Oxford University Press. https://doi.org/10.1093/acprof:oso/9780199241323.001.0001.

Corona Uhink, G. (1964). Palabras del Dr. Guillermo Corona Uhink Prof. de Medicina Humanistica de la Escuela Nacional de Medicina de la U.N.A.M. *Estudios Sobre Esterilidad, 15*(2), 105–116.

Corro, S. (1985, September 29). A los 22 años murio el Centro Médico más avanzado de América Latina. *Proceso.* Retrieved from http://www.proceso.

com.mx/142125/a-los-22-anos-murio-el-centro-medico-mas-avanzado-de-america-latina.

Crisp, R. (1987). Persuasive advertising, autonomy, and the creation of desire. *Journal of Business Ethics, 6*(5), 413–418. https://doi.org/10.1007/BF00382898.

Cruz-Martínez, A. (2003, May 23). Cumple 15 años el primer caso de nacimiento in vitro en México. *La Jornada.* Section: Sociedad y Justicia.

Cueno Pareto, S., & Gaviño Gaviño, F. (2015). Alfonso Guitiérrez Najár. *Reproducción, 7,* 138–140.

Davis-Floyd, R., & Dumit, J. (1998). *Cyborg babies: From Techno-sex to Technotots.* Great Britain: Routledge.

de Barbieri, T. (2000). Derechos reproductivos y sexuales. Encrucijada en tiempos distintos. *Revista Mexicana de Sociología, 62*(1), 45. https://doi.org/10.2307/3541178.

de la Cadena, M., Lien, M. E., Blaser, M., Jensen, C. B., Lea, T., Morita, A., ... Wiener, M. (2015). Anthropology and STS: Generative interfaces, multiple locations. *HAU: Journal of Ethnographic Theory, 5*(1), 437–475. https://doi.org/10.14318/hau5.1.020.

de la Torre, R. (2008). La Iglesia Católica en el México contemporáneo. Resultados de una prueba de contraste entre jerarquía y creyentes. *L'Ordinaire des Amériques, 210,* 27–46. https://doi.org/10.4000/orda.2616.

de Márquez, V. B. (1984). El proceso social en la formación de políticas/ el caso de la planificación familiar en México. *Estudios Sociológicos, 11*(2–3), 309–333.

De Saille, S. (2017). Knowledge *as resistance: The feminist international network of resistance to reproductive and genetic engineering.* London: Palgrave Macmillan (Imprint).

Dimond, R., & Stephens, N. (2018). Legalising *mitochondrial donation: Enacting ethical futures in UK biomedical politics.* Printforce, The Netherlands: Palgrave Macmillan. ISBN 978-3-319-74644-9, 978-3-319-74645-6, p. 147.

Dyer, A. R. (1997). Ethics, advertising, and assisted reproduction: The goals and methods of advertising. *Women's Health Issues, 7*(3), 143–148.

Ehrlich, P. R. (1968). *The population bomb.* New York: Ballantine Books.

El incremento demográfico y la planeación familiar. (1964, Mayo–Ago). *Estudios Sobre Esterilidad, XV*(2).

Eliot, T. S. (1938). *Old Possum's book of practical cats.* New York: Harcourt, Brace & Co.

Espinosa de los Reyes, V. (1981). El doctor Isidro Espinosa de los Reyes y los inicios de la atención materno-infantil en México. *Voces Académicas, 117*(2), 81–87.

Espinosa de los Reyes Sánchez, V. M. (2008). Datos biográficos del Dr. Isidro Espinosa de los Reyes. *Boletín Mexicano de Historia y Filosofía de La Medicina, 11*(2), 64–67.

Espinosa de los Reyes Sánchez, V. M. (2016). La asistencia materno-infantil en México entre 1921 y 1930 por parte del Departamento de Salubridad Pública. *Gaceta Médica de México, 152*, 231–245.

Fajardo-Ortiz, G. (2014). Setenta años de medicina en el Instituto Mexicano del Seguro Social. *Revista Médica Del Instituto Mexicano del Seguro Social, 52*(2), 228–231.

Fajardo-Ortiz, G. (2015). Cuatro etapas en la historia del Centro Médico Nacional Siglo XXI del IMSS. *Revista Médica Del Instituto Mexicano Del Seguro Social, 53*(5), 656–663.

Fajardo Ortiz, G., & Sánchez González, J. M. (2005). La medicina mexicana de 1901 a 2003. *Revista Latinoamericana de Patología Clínica y Medicina de Laboratorio, 52*(2), 118–123.

Fassin, D. (2009). Another politics of life is possible. *Theory, Culture & Society, 26*(5), 44–60. https://doi.org/10.1177/0263276409106349.

Finkler, K. (1994). *Women in pain: Gender and morbidity in Mexico.* Philadelphia: University of Pennsylvania Press.

Finkler, K. (2004). Biomedicine globalized and localized: Western medical practices in an outpatient clinic of a Mexican hospital. *Social Science & Medicine, 59*(10), 2037–2051.

Flores Palacios, M. L., & Sánchez Santana, A. G. (2006). Capítulo 11. Estereotipos demográficos y ocupacionales de la mujer y el hombre en la televisión mexicana. In *Anuario de Investigación de la Comunicación CONEICC XIII* (pp. 257–271). Retrieved from https://www.researchgate.net/profile/Ana_Gaspar4/publication/27392963_Rehacer_el_tejido_de_Penelope_mujeres_y_reproduccion_de_la_emigracion/links/53ee50850c-f23733e80c528e.pdf#page=257.

Folch-Lyon, E., de la Macorra, L., & Bruce Schearer, S. (1981). Focus group and survey research on family planning in Mexico. *Studies in Family Planning, 12*(12), 409–432.

Franco Migues, D. (2012). Ciudadanos de ficción: Discursos y derechos ciudadanos en las telenovelas mexicanas. *El caso Alma de Hierro. Comunicación y Sociedad, Nueva época* (17), 41–71.

Franklin, S. (1997). *Embodied progress: A cultural account of assisted conception.* London: Routledge.

Franklin, S. (2013). *Biological relatives: IVF, stem cells, and the future of kinship.* Durham: Duke University Press.

Franklin, S. (2014). Analogic return: The reproductive life of conceptuality. *Theory, Culture & Society, 31*(2–3), 243–261. https://doi.org/10.1177/0263276413510953.

Franklin, S., & Inhorn, M. C. (2016). Introduction. *Reproductive Biomedicine & Society Online, 2,* 1–7. https://doi.org/10.1016/j.rbms.2016.05.001.

Franklin, S., Lury, C., & Stacey, J. (2000). *Global nature, global culture.* London: Sage.

Freidson, E. (1983). Viewpoint: Sociology and medicine—A polemic. *Sociology of Health and Illness, 5*(2), 208–219. https://doi.org/10.1111/1467-9566.ep10491530.

Freidson, E. (1988). *Profession of medicine: A study of the sociology of applied knowledge.* Chicago: University of Chicago Press.

Frenk, J. (2002). *Los institutos Nacionales de Salud de Mexico.* México: Secretaria de Salud.

Fuentes, C. (1993, December 2). Tribuna/TLC, el día siguiente. *El País.* Retrieved from https://elpais.com/diario/1993/12/02/opinion/754786811_850215.html.

Galache Vega, P., Hernandez Ayup, S., Arenas Santillán Gómez, A., Ruy Sánchez Aguilar, M., &Santos Haliscak, R. (1999). Fertilización in Vitro. In E. Vázquez Benítez (Ed.), *Medicina Reproductiva* (1st ed., p. 279, Chap. 54). México: JGH Editores. ISBN 9706810250.

Galache Vega, P., Santos Haliscak, R., Batiza, R., Saucedo de la Llata, E., Arenas Montezco, L., García Villafana, G., & Ayup Hernández, S. (2003). Fertilización in Vitro. In E. Vázquez Benítez (Ed.), *Medicina Reproductiva* (2°). México, D.F., México: Editorial Manual Moderno, AMMR.

Gallo, D. (1960). Frigidez, Orgasmo y Esterilidad. *Estudios sobre Esterilidad, 11*(3), 151.

Garza-Gordoa, M. (2012). Imagen visual del símbolo del Instituto Nacional de Perinatología Isidro Espinosa de los Reyes. *Perinatología y Reproducción Humana, 26*(2), 133–137.

Geels, F. W. (2007). Transformations of large technical systems: A multilevel analysis of the Dutch highway system (1950–2000). *Science, Technology, & Human Values, 32*(2), 123–149. https://doi.org/10.1177/0162243906293883.

Gerrits, T. (2016). Assisted reproductive technologies in Ghana: Transnational undertakings, local practices and 'more affordable' IVF. *Reproductive*

Biomedicine & Society Online, 2, 32–38. https://doi.org/10.1016/j.rbms. 2016.05.002.

Giner, J., Merino, G., Luna, J., & Aznar, R. (1974). Evaluation of the Sims-Huhner Postcoital Test in Fertile Couples*(*Supported in part by a Ford Foundation grant). *Fertility and Sterility, 25*(2), 145–148. https://doi. org/10.1016/S0015-0282(16)40213-X.

Ginsburg, F. D., & Rapp, R. (Eds.). (1995). *Conceiving the new world order: The global politics of reproduction* (p. xii). Berkeley: University of California Press (Cloth and Paper).

Gire, G. de I. en R. E. A., & Radar 4. (2015). *Niñas y Mujeres sin Justicia. Derechos Reproductivos en México.*

González de Bustamante, C. (2009). Dependency and development: The importance of TV news in the history of Mexican television. *Galáxia* (18), 247–262.

González de Bustamante, C. (2010). 1968 olympic dreams and tlatelolco nightmares: Imagining and imaging modernity on television. *Mexican Studies/Estudios Mexicanos, 26*(1), 1–30. https://doi.org/10.1525/ msem.2010.26.1.1.

González-Gutierrez, J. T. (1962). La adopción en matrimonios definitivamente estériles. *Estudios sobre Esterilidad, 13*(3), 137–146.

González-Santos, S. P. (2011a). Space, structure and social dynamics within the clinical setting: Two case studies of assisted reproduction in Mexico City. *Health & Place, 17*(1), 166–174. https://doi.org/10.1016/j. healthplace.2010.09.013.

González-Santos, S. P. (2011b). *The sociocultural aspects of assisted reproduction in Mexico* (Doctoral, University of Sussex). Retrieved from http://sro.sussex. ac.uk/7081/.

González-Santos, S. P. (2014). Specialization in action the genealogy and current state of assisted reproduction. *Bulletin of Science, Technology & Society, 34*(1–2), 33–42. https://doi.org/10.1177/0270467614538948.

González-Santos, S. P. (2016a). From esterilología to reproductive biology: The story of the Mexican assisted reproduction business. *Reproductive Biomedicine & Society Online, 2*, 116–127. https://doi.org/10.1016/j. rbms.2016.10.002.

González-Santos, S. P. (2016b). Peregrinar: el ritual de la reproducción asistida. In C. Straw, E. Vargas, M. Viera Cherro, & M. Tamanini (Eds.), *Rerpodução Assistida: e relações de género na América liana.* Curitiba, Braisl: Ed. CRV.

González-Santos, S. P. (2017). Shifting the focus in the legal analysis of the first MST case. *Journal of Law and the Biosciences, 1*(6). https://doi.org/10.1093/jlb/lsx022.

González Santos, S. P., Stephens, N., & Dimond, R. (2018). Narrating the first "three-parent baby": The initial press reactions from the United Kingdom, the United States, and Mexico. *Science Communication, 40*(4), 419–441. https://doi.org/10.1177/1075547018772312.

Goode, W. J. (1960). The profession: Reports and opinion. Encroachment, charlatanism, and the emerging professions: Psychology medicine and sociology. *American Sociological Review, 25*(6), 902–914. Retrieved from http://www.jstor.org/stable/2089988.

Gorodovsky, J., & Rice-Wray, E. (1968, Septiembre-diciembre). Nuevo método en el control de la fertilidad. Inyección mensual de depósito. *Estudios sobre Esterilidad, XIX*(3), 154.

Gray, M. (2018, March 28). *More than 4,000 eggs and embryos lost in Cleveland fertility clinic tank failure.* Retrieved February 15, 2019, from CNN website: https://edition.cnn.com/2018/03/27/health/clevelandfertility-clinic-eggs-embryos-lost/index.html.

Gual-Castro, C. (2000). Programas universitarios de posgrado en biología de la reproducción. *Gaceta Médica de México, 136*(3), S75–S78.

Gual-Castro, C. (2012). El inicio del embarazo en la mujer, la planiicación familiar y el uso de anticonceptivos* Conferencia "Dr. Eduardo Liceaga." *Revista Médica Del Hospital General de México. Méx, 75*(4), 238–246.

Guerrero, D. C. (1953). Progesos y Problemas a Resolver en la Esterilidad. *Estudios Sobre Esterilidad, 4*(4), 198.

Guerrero, D. C. (1954). Aspectos Sociales de la Esterilidad. *Estudios Sobre Esterilidad, 5*(1), 29.

Guerrero, D. C. (1964). Management of the infertile couple: 25 years' experience. *Fertility and Sterility, 15*(5), 534–542. https://doi.org/10.1016/S0015-0282(16)35348-1.

Guerrero, D. C., & Álvarez Bravo, A. (1956, Mayo–Diciembre). *Impresiones del Segundo Congreso Mundial de Fertilidad y Esterilidad, VII*(3), 103.

Gürtin, Z. B. (2016). Patriarchal pronatalism: Islam, secularism and the conjugal confines of Turkey's IVF boom. *Reproductive BioMedicine and Society Online, 2*, 39–46. https://doi.org/10.1016/j.rbms.2016.04.005.

Gutierrez Najar, A. (2003a). Balstocisto: un nuevo horizonte. Sección XIII La Reproducción Asistida. Cap. 54 (p. 391). In E. Vázquez Benítez (Ed.), *Medicina Reproductiva* (2°). México, D.F., Mexico: Editorial Manual Moderno, AMMR.

Gutierrez Najar, A. (2003b). Resultados de la reproducción asistida en Latinoamérica. Sección XIII La Reproducción Asistida. Cap. 57 (p. 421). In E. Vázquez Benítez (Ed.), *Medicina Reproductiva* (2°). México, D.F., Mexico: Editorial Manual Moderno, AMMR.

Gutierrez Najar, A., Navarro, M. C., Martin, O. E., & Diaz, M. (1992). Fertilización in vitro-GIFT: evaluación de una clínica 1988–1992. *Ginecología y Obstetricia de México, 60*(S1), 100.

Gutiérrez-Sánchez, S. (2000). Transición de la alta a la baja fecundidad en México. Cuadernos de investigación, Cuarta Época, No. 12. México, Universidad Autónoma del Estado de México.

Gutmann, M. (2009). Planning men out of family planning: A case study. *Sexualidad, Salud y Sociedad - Revista Latinoamericana, 1*, 104–124.

Hables Gray, C. (2002). *Cyborg citizen.* New York and London: Routledge.

Haraway, D. J. (2015). *Simians, cyborgs, and women: The reinvention of nature.* New York: Routledge.

Haraway, D. J. (2008). *When species meet.* Minneapolis: University of Minnesota Press.

Haraway, D. J. (2016). *Staying with the trouble: Making kin in the Chthulucene.* Durham: Duke University Press.

Hawkins, J. (2013). Selling art: An empirical assessment of advertising on fertility clinics' websites. *Indiana Law Journal, 88*(4), 1148–1179.

Hawkins, J. (2017). Exploiting advertising. *Law & Contemporary Problems, 80*(3), 43–71.

Hedgecoe, A., & Martin, P. (2003). The drugs don't work: Expectations and the shaping of pharmacogenetics. *Social Studies of Science, 33*(3), 327–364. https://doi.org/10.1177/03063127030333002.

Hernández Ayup, S., Galache Vega, P., Santillán Gómez, A., Ruy Sánchez Aguilar, M., & Santos Haliscak, R. (1999). Fertilización in Vitro. Cap. 54 (p. 279). In E. Vázquez Benítez (Ed.), *Medicina Reproductiva en México* (Primera). JGH Editores S.A. de C.V.

Hernández Ayup, S., Santos Haliscak, R., García Martínez, M., Morales Caballero, F., Loret de Mola Gutiérrez, R., & Galache Vega, P. (1989). Gift: Reproductive reality for the sterile couple. *Ginecología Y Obstetricia De México, 57*, 315–319.

Hernández Valencia, M., & Saucedo García, R. (2018). Contribuciones del IMSS a la medicina mundial. In T. Miguel Ortega, J. de J. Arriaga Dávila, A. C. Sepúlveda Vildósola, F. A. Salamanca Gómez, & C. Martínez Castuera Gómez (Eds.), *Pasado, presente y futuro.* Instituto Mexicano del Seguro Social. IPN.

Hörbst, V. (2016). 'You cannot do IVF in Africa as in Europe': The making of IVF in Mali and Uganda. *Reproductive Biomedicine & Society Online, 2*, 108–115. https://doi.org/10.1016/j.rbms.2016.07.003.

Huang, J. Y. J., Discepola, F., Al-Fozan, H., & Tulandi, T. (2005). Quality of fertility clinic websites. *Fertility and Sterility, 83*(3), 538–544. https://doi.org/10.1016/j.fertnstert.2004.08.036.

Hughes, T. P. (1979). The electrification of america: The system builders. *Technology and Culture, 20*(1), 124. https://doi.org/10.2307/3103115.

Hughes, T. P. (1986). The seamless web: Technology, science, etcetera, etcetera. *Social Studies of Science, 16*(2), 281–292.

Hughes, T. P. (1987). The evolution of large technological systems. In W. E. Bijker, T. P. Hughes, & T. Pinch (Eds.). *The Social construction of technological systems: New directions in the sociology and history of technology.* Cambridge, MA: MIT Press.

Hughes, T., & Coutard, O. (1996). Fifteen years of social and historical research on large technical systems. An interview with Thomas Hughes. *FLUX Cahiers scientifiques internationaux Réseaux et Territoires, 12*(25), 44–47.

Inhorn, M. C., & Balen, F. (2002). *Infertility around the globe: New thinking on childlessness, gender and reproductive technologies.* Berkeley: University of California Press.

Inhorn, M. C., & Birenbaum-Carmeli, D. (2008). Assisted reproductive technologies and culture change. *Annual Review of Anthropology, 37*(1), 177–196. https://doi.org/10.1146/annurev.anthro.37.081407.085230.

Jones, H. W. (2003). IVF: Past and future. *Reproductive BioMedicine Online, 6*(3), 375–381. https://doi.org/10.1016/S1472-6483(10)61860-3.

Kably Ambe, A. (1988). Fertilización in Vitro. In J. Delgado Urdapilleta & A. Kably Ambe (Eds.), *V Reunión Anual. Experiencia en México* (Conference Report) (pp. 111–124). México: Instituto Nacional de Perinatología.

Kably Ambe, A., Salazar López Ortiz, C., Serviere Zaragoza, C., Velázquez Cornejo, G., Pérez Peña, E., Santos Haliscack, R., … Gaviño Gaviño, F. (2012). Consenso Nacional Mexicano de Reproducción Asistida. *Ginecología Y Obstetricia De México, 80*(9), 581–624.

Kably Ambe, A., et al. (1992). *Ginecología y Obstetrica de México, 60*(Suppl 1), 93.

Karchmer, S. (1989). *Temas Selectos en Repoducción Humana Parte I.* Mexico: Instituto Nacional de Perinatología.

Karchmer, S. (1999). *Instituto Nacional de Perinatología 1983–1993* (Libro Negro). México: INPer.

Kelly, J. J. (1994). Article 27 and mexican land reform: The legacy of Zapata's dream. *Human Rights Literature Review, 25*(541), 541–570.

Krause, E. L., & De Zordo, S. (2012). Introduction ethnography and biopolitics: Tracing 'rationalities' of reproduction across the north–south divide. *Anthropology & Medicine, 19*(2), 137–151. https://doi.org/10.1080/13648 470.2012.675050.

Kunhardt Rasch, J. (1999). Presentación (p. 1). In E. Vázquez Benítez (Ed.), *Medicina Reproductiva en México* (Primera). Mexico: JGH Editores S.A. de C.V.

Larson, M. S. (1977). *The rise of professionalism: A sociological analysis.* Berkeley: University of California Press.

Latour, B. (2011). Introduction: How to handle technical innovation. Course, MOOC on *Scientific Humanities.* Retrieved from http://www.bruno-latour. fr/node/532.

Law, J. (1992). Notes on the theory of the actor-network: Ordering, strategy, and heterogeneity. *Systemic Practice, 5*(4), 379–393.

Law, J. (2009). Actor network theory and material semiotics. *The New Blackwell Companion to Social Theory, 3,* 141–158.

Law, J., & Mol, A. (1995). Notes on materiality and sociality. *Sociological Review, 43*(2), 274–294. https://doi.org/10.1111/1467-954X.ep9505171482.

Leigh Star, S. (1999). The ethnography of infrastructure. *American Behavioral Scientist, 43*(3), 377–391. https://doi.org/10.1177/00027649921955326.

León Olivares, F. (1999). *Syntex, origen, apogeo y perdida de una industria estrategica para Mexico* (MSc dissertation). Instituto Politecnico Nacional.

León Olivares, F. (2001). El origen de Syntex, una enseñanza histórica en el contexto de ciencia, tecnología y sociedad. *Revista de la Sociedad de Química de México, 45*(2), 4.

Lerner, S. (1967). La investigación y la planeación demográficas en México. *Estudios Demográficos y Urbanos, 1*(1), 9. https://doi.org/10.24201/edu. v1i01.27.

Lewkowicz, E. (2014). Rebel love: Transnational teen TV vs. Mexican telenovela tradition. *Continuum: Journal of Media & Cultural Studies, 28*(2), 265–280. https://doi.org/10.1080/10304312.2013.854869.

Lewkowicz, E. (2015). Cinderella's having a ball: Humoring Mexico's "ugly" TV formula. *Feminist Media Studies, 15*(2), 271–286. https://doi.org/10.10 80/14680777.2014.924980.

Lock, M., & Kaufert, P. A. (2006). *Pragmatic women and body politics.* Cambridge: Cambridge University Press.

López-Llera, M. (2003, February 27). *Pasajes inéditos en la vida y muerte del hospital de gineco-obstetricia del Centro Médico Nacional.* Retrieved from http://webcache.googleusercontent.com/search?q=cache:vGRZh21DxZ8J: gineco2.com.mx/fotos_archivos/Pasajeshgo2.doc+&cd=1&hl=en&ct= clnk&gl=mx&client=firefox-b-d.

Lozano, J. L. (1952, Julio). El factor psico-somático en la infertilidad de la mujer. *Estudios Sobre Esterilidad, III*(3), 153.

MacKenzie, D. (2012). Missile accuracy: A case study in the social processes of technological change. In W. E. Bijker, T. P. Hughes, & T. Pinch (Eds.), *The social construction of technological systems: New directions in the sociology and history of technology* (Anniversary ed.). Cambridge, MA: MIT Press.

Madeira, J. L. (2013). Selling art or selling out?: A response to selling art— An empirical assessment of advertising on fertility clinics' websites. *Indiana Law Journal, 88*(4), 1181–1185.

Maher, J. (2014). Something else besides a father: Reproductive technology in recent Hollywood film. *Feminist Media Studies, 14*(5), 853–867. https://doi.org/10.1080/14680777.2013.831369.

Maldonado, L. D. J. (2017). *Análisis del embarazo adolescente en México 2000–2012 (Tesina).* Centro de Investigación y Docencia Económicas, A.C. Retrieved from http://repositorio-digital.cide.edu/bitstream/handle/11651/2243/158721.pdf?sequence=1&isAllowed=y.

Martínez, D. P. (1964, Mayo–Ago). El incremento demográfico y la planeación familiar. *Estudios Sobre Esterilidad, XV*(2), 101.

Martínez Manaton, J. (1994). Veinte años de política de población en México. *Papeles de Población*, núm. 4–5, julio–octubre, pp. 59–62. Universidad Autónoma del Estado de México, Toluca, México.

Martínez Manautou, J. (2013). Departamento de Investigación Científica del IMSS Su inicio en 1966. In F. A. Salamanca Gómez, J. A. González Anaya, F. Cruz Vega, & F. Aceves Ávila (Eds.), *Investigación en salud*. México, D.F.: Editorial Alfil: Gobierno de la República: IMSS: Academia Mexicana de Cirugía, A.C.: Fundación IMSS.

Martinez-Manautou, J., Cortez, V., Giner, J., Aznar, R., Casasola, J., & Rudel, H. W. (1966). Low doses of progestogen as an approach to fertility control. *Fertility and Sterility, 17*(1), 49–58. https://doi.org/10.1016/S0015-0282(16)35825-3.

Martinez-Manautou, J., Giner-Velasquez, J., Cortes-Gallegos, V., Aznar, R., Rojas, B., Guitterez-Najar, A., & Rudel, H. W. (1967). Daily progestogen for contraception: A clinical study. *British Medical Journal, 2*(5554), 730–732. https://doi.org/10.1136/bmj.2.5554.730.

Mateos Fournier, M. (1964). El incremento demográfico y la planeación familiar. Symposium. *Estudios Sobre Esterilidad, 15*(2), 60–64.

Mateos, F., Román, A., Valdéz, A., & Pineda, R., (1969). *Nuevas aportaciones para la evaluación del problema del aborto criminal en México* (p. 29). México: Fundación para Estudios de la Población.

Mateos, G., & Suárez-Díaz, E. (2014). Peaceful atoms in Mexico. In E. Medina, I. C. Marques, & C. Holmes (Eds.), *Beyond imported magic: Essays on science, technology, and society in Latin America*. Cambridge, MA: The MIT Press.

McNeil, M., Varcoe, I., & Yearley, S. (1990). *The new reproductive technologies. no. 28 de explorations in sociology*. British Sociological Association. Conference. St. Martin's Press, 1990. ISBN 0312035993, 9780312035990. p. 257.

Mejía-Modesto, A. (2007). *Poliíticas de población, los derechos humanos y la individualización*. México: Gobierno del Estado de México, Consejo Estatal de Población.

Mendoza García, M. E., & Tapia Colocia, G. (2010). La situacion demografica de Mexico. *Situacion Demografica de Mexico 1910–2010*. UNFPA. Fondo de Población de las Naciones Unidas.

Michelle, C. (2007). "Human clones talk about their lives": Media representations of assisted reproductive and biogenetic technologies. *Media, Culture & Society, 29*(4), 639–663. https://doi.org/10.1177/0163443707078425.

Mier y Terán, M. (1991). El gran cambio demográfico. *Demos, 4*, 4–5.

Moghissi, K. S. (1976). Postcoital test: Physiologic basis, technique, and interpretation. *Fertility and Sterility, 27*(2), 117–129. https://doi.org/10.1016/S0015-0282(16)41648-1.

Mohr, S., & Koch, L. (2016). Transforing social contracts: The social and cultural history of IVF in Denmark. *Reproductive Biomedicine & Society Online, 2*, 88–96. https://doi.org/10.1016/j.rbms.2016.09.001.

Mol, A. (2002). *The body multiple: Ontology in medical practice*. Durham: Duke University Press.

Morales Suárez, M. (1999). Princesa 3 pedernal serpiente emplumada simbolo del INPer. *Perinatología y Reproducción Humana, 13*(4), 255–263.

Morales Suárez, M. (2010). Trayectoria del Dr. Eduardo Jurado García (1921–1998). Un acercamiento a su vida y obra. *Perinatología y Reproducción Humana, 24*(3), 207–211.

Morán Villota, C., Arechavaleta Velasco, F. J., & Maya Núñez, G. (2013). Biología de la reproducción. In F. A. Salamanca Gómez, J. A. González Anaya, F. Cruz Vega, & F. Aceves Ávila (Eds.), *Investigación en salud*. México, D.F.: Editorial Alfil: Gobierno de la República: IMSS: Academia Mexicana de Cirugía, A.C.: Fundación IMSS.

Moreira, T., & Palladino, P. (2005). Between truth and hope: On Parkinson's disease, neurotransplantation and the production of the 'self'. *History of the Human Sciences, 18*(3), 55–82. https://doi.org/10.1177/0952695105059306.

Morgan, L. M., & Roberts, E. F. S. (2012). Reproductive governance in Latin America. *Anthropology & Medicine, 19*(2), 241–254. https://doi.org/10.108 0/13648470.2012.675046.

Morton, A. D. (2003). Structural change and neoliberalism in Mexico: "Passive revolution" in the global political economy. *Third World Quarterly, 24*(4), 631–653.

Mulholland, J., Mallidis, C., Agbaje, I., & McClure, N. (2011). Male diabetes mellitus and assisted reproduction treatment outcome. *Reproductive BioMedicine Online, 22*(2), 215–219. https://doi.org/10.1016/j.rbmo.2010.10.005.

Mulkay, M. (1994a). Science and family in the great embryo debate. *Sociology, 28*(3), 699–715. https://doi.org/10.1177/0038038594028003004.

Mulkay, M. (1994b). The Triumph of the pre-embryo: Interpretations of the human embryo in parliamentary debate over embryo research. *Social Studies of Science, 24*(4), 611–639.

Murillo Gutiérrez, E., & Valdez la Vallina, F. (1953, Abril). Factores etiológicos de esterilidada en 100 parejas. *Estudios sobre Esterilidad, IV*(2), 90.

Murphy, M. (2017). *The economization of life*. Durham and London: Duke University Press.

Nagel, J. S. (1978). Mexico's population policy turnaround. *Population Bulletin* (Population Reference Bureau Inc.), *33*(5), 1.

Najam, A. (1996). A developing countries' perspective on population, environment, and development. *Population Research and Policy Review, 15*, 1–9.

Napolitano, V. (2009). The Virgin of Guadalupe: A nexus of affect. La Virgen de Guadalupe: Un Point de Fusion Des Affects. *Journal of Royal Anthropological Institute, 15*(1), 96–112. https://doi.org/10.1111/j.1467-9655.2008.01532.x.

Nazar-Beutelspacher, A., Zapata-Martelo, E., & Vázquez-García, V. (2004). Population policies and women's nutrition: A study on six rural communities in Chiapas. *México Agricultura Sociedad y Desarrollo, 1*(2), 147–162.

Nerlich, B., Johnson, S., & Clarje, D. D. (2003). The first 'designer baby': The role of narratives, cliche's and metaphors in the year 2000 media debate. *Science as Culture, 12*(4), 471–498. https://doi.org/10.1080/0950543032000150328.

Neyer, G., & Bernardi, L. (2013, August). *Feminist perspectives on motherhood and assisted reproduction*. Paper presented at the XXVII International IUSSP Conference, Busan, Korea. Retrieved from http://iussp.org/sites/default/files/event_call_for_papers/IUSSP%20Neyer%20Bernadi%20-%20Reproduction%20updated.pdf.

O'Hara, H. (1944, June). Review of: Doña Eugenesia y Otros Personajes. *American Journal of Public Health, 34*(6), 662.

Olavarría, M. E. (2018). *La gestación para otros. Parentesco, tecnología y poder*. México: Universidad Autónoma Metropolitana and Gedisa.

Olavarría Patiño, M. E. (2018). La gestante sustituta en México y la noción de trabajo reproductivo. *Revista Interdisciplinaria de Estudios de Género de El Colegio de México, 4*, 1–31.

Orozco Gómez, G. (2006). La telenovela en México: ¿de una expresión cultural a un simple producto para la mercadotecnia? *Comunicación y Sociedad* (0188–252X)(6), 11–35.

Ortega, M., Arriaga Dávila, J. de J., Sepúlveda Vildósola, A. C., Salamanca Gómez, F. A., & Martínez Castuera Gómez, C. (Eds.). (2018). Contribuciones del IMSS a la medicina mundial. *Pasado, presente y futuro*. Instituto Mexicano del Seguro Social. IPN. Retrieved from https://www.ipn.mx/assets/files/secgeneral/docs/SG/Libro%20IMSS%2075%20a%C3%B1os.pdf.

Ortiz Pinchetti, F. (1984, Marzo 17). Controversia entre médicos. *Proceso* (385). https://www.proceso.com.mx/.

Osborne-Thompson, H. (2014). Seriality and assisted reproductive technologies in celebrity reality television. *Feminist Media Studies, 14*(5), 877–880. https://doi.org/10.1080/14680777.2014.952876.

Oudshoorn, N. (1993). United we stand: The pharmaceutical industry, laboratory, and clinic in the development of sex hormones into scientific drugs, 1920–1940. *Science, Technology & Human Values, 18*(1), 5–24.

Oudshoorn, N. (1997). From population control politics to chemicals: The WHO as an intermediary organization in contraceptive development. *Social Studies of Science, 27*(1), 41–72.

Oudshoorn, N. (1999). On masculinities, technologies, and pain the testing of male contraceptives in the clinic and the media. *Science, Technology, & Human Values, 24*(2), 265–289.

Oudshoorn, N., & Pinch, T. (Eds.). (2003). *How users matter: The co-construction of users and technologies*. Cambridge, MA: MIT Press.

Pastor, M. (2010). El marianismo en México: una mirada a su larga duración. *Cuicuilco, 17*(48), 257–277.

Paz, S. (2015, February 5). *El mexicano que detonó la revolución sexual*. Retrieved October 29, 2018, from http://www.conacytprensa.mx/index.php/sociedad/personajes/154-el-mexicano-que-detono-la-revolucion-sexual.

Pearson, R. C. (2005). Fact or Fiction? Narrative and reality in the Mexican telenovela. *Television & New Media, 6*(4), 400–406. https://doi.org/10.1177/1527476405279863.

Pérez-Bustos, T. (2017). "No es sólo una cuestión de lenguaje": lo inaudible de los estudios feministas latino-americanos en el mundo académico anglosajón. *Scientiae Studia, 15*(1), 59. https://doi.org/10.11606/51678-31662017000100004.

Pérez, P., & Aguirre, U. (1989). In G. Soberón, J. Kumate, & J. Laguna (compliladores). Cuahutemoc Valdés (coordinador). *La Salud en México: Testimonios 1988. Especialidades Médicas en México.*

Pérez Peña, E. (1999). Introducción: La Reproducción Asistida. Cap. 46 (p. 249). In E. Vázquez Benítez (Ed.), *Medicina Reproductiva en México* (Primera). México: JGH Editores S.A. de C.V.

Pérez Peña, E. (2003). Sección XIII: Reproducción Asistida. In E. Vázquez Benítez (Ed.), *Medicina Reproductiva* (2°). México, D.F., México: Editorial Manual Moderno, AMMR.

Pérez Peña, E., Gutierrez Gutierrez, A., Garza Morales, A., Velez, P., Rojas, R. F., Gutierrez, T., & Canul, R. M. (2003). Reproducción Asistida. Alcances y Limitaciones. Sección XIII: Reproducción Asistida. Cap. 38 (p. 275). In E. Vázquez Benítez (Ed.), *Medicina Reproductiva* (2°). México, D.F., Mexico: Editorial Manual Moderno, AMMR.

Phelan, J. (1996). *Narrative as rhetoric: Technique, audiences, ethics, ideology.* Columbus: Ohio State University Press.

Pick de Weiss, S. (1987). Actitudes, conocimientos y conductas de planificacion familiar en México: una década de investigación psicosocial (1975–1985). *Revista Mexicana de Psicología, 3*(2), 155–160.

Pinch, T. J., & Bijker, W. E. (1984). The social construction of facts and artefacts: Or how the sociology of science and the sociology of technology might benefit each other. *Social Studies of Science, 14*(3), 399–441. https://doi.org/10.1177/030631284014003004.

Pope Pius XII. (1956). Discurso de S.S. Pio XII a los Miembros del Segundo Congreso Mundial de Fertilidad y Esterilidad. *Estudios Sobre Esterilidad, 7*(2–3).

Prados, F. J., Debrock, S., Lemmen, J. G., & Agerholm, I. (2012). The cleavage stage embryo. *Human Reproduction, 27*(suppl 1), i50–i71. https://doi.org/10.1093/humrep/des224.

Preston, J., & Dillon, S. (2004). *Opening Mexico: The making of a democracy.* New York: Farr, Straus, and Giroux.

Quark Adams, A. (2019). Outosorcing regulatory decision-making: "International" epistemic communities, transnational firms, and pesticides residue standards in India. *Science Technology & Human Values, 44*(1), 3–28. https://doi.org/10.1177/0162243918779123.

Rabell, E. (1964). Palabras del profesor y licenciado Enrique Rabell del consejo nacional técnico de la educación de la Secretaría de Educación Pública. La explosión demográfica y la acción educativa del régimen. *Estudios Sobre Esterilidad, 15*(2), 77–80.

Rabinow, P., & Rose, N. (2006). Biopower today. *BioSocieties, 1*(2), 195–217. https://doi.org/10.1017/S1745855206040014.

Radin, M. J. (1987). Market-inalienability. *Harvard Law Review, 100*(8), 1849. https://doi.org/10.2307/1341192.

Raju, G. A. R., Prakash, G. J., Krishna, K. M., Madan, K., Narayana, T. S., & Krishna, C. H. R. (2012). Noninsulin-dependent diabetes mellitus: Effects on sperm morphological and functional characteristics, nuclear DNA integrity and outcome of assisted reproductive technique. *Andrologia, 44*(s1), 490–498. https://doi.org/10.1111/j.1439-0272.2011.01213.x.

Ramírez, C. S. (1962). *Esterilidad y fruto: Psicología de la función procreativa.* México: Pax-México.

Reinehart, W. (1977). *La minipíldora. Una alternativa limitada para ciertas mujeres* (Anticonceptivos Orales No. 3) (pp. 61–79). Universidad de George Washington. Retrieved from https://www.k4health.org/sites/default/files/751074SPA.PDF.

Reinhold, R. (1979, November 5). Mexico's birth rate seems off sharply. *The New York Times.* Retrieved from The New York Times Archive.

Rice Wray, E. (1970, Enero–Abril). Reproducción después del uso de anticonceptivos. *Estudios sobre Esterilidad, XXI*(1), 30.

Roberts, E. F. S. (2006). God's laboratory: Religious rationalities and modernity in Ecuadorian in vitro fertilization. *Culture, Medicine and Psychiatry, 30*(4), 507–536. https://doi.org/10.1007/s11013-006-9037-8.

Roberts, E. F. S. (2007). Extra embryos: The ethics of cryopreservation in Ecuador and elsewhere. *American Ethnologist, 34*(1), 181–199.

Roberts, E. F. S. (2016). Resources and race: Assisted reproduction in Ecuador. *Reproductive Biomedicine & Society Online, 2,* 47–53. https://doi.org/10.1016/j.rbms.2016.06.003.

Rodriguez Argüelles, J. (1999). Prevención de la responsabilidad legal en relación con la reproducción. In E. Vázquez Benítez (Ed.), *Medicina Reproductiva en México* (Primera). Mexico: JGH Editores S.A. de C.V.

Rodriguez Argüelles, J. (2003). Prevención de la responsabilidad legal en relación con la reproducción. Sección XVII: Aspectos legales de los estudios y tratamientos relacionados con la reproducción. Cap. 72 (p. 539). In E. Vázquez Benítez (Ed.), *Medicina Reproductiva* (2º). México, D.F., Mexico: Editorial Manual Moderno, AMMR.

Rodriguez Medina, L. (2019). A geopolitics of bad English. *Tapuya: Latin American Science, Technology and Society, 2*(1), 1–7. https://doi.org/10.1080/25729861.2019.1558806.

Rogers, E. M. (1983). *Diffusion of innovations* (3rd ed.). New York and London: Free Press and Collier Macmillan.

Ramirez, S. (1960, Septiembre–diciembre). Factores Culturales en la Esterilidad e Infertilidad. *Estudios Sobre Esterilidad, XI*(3), 146.

Saade Granados, M. (2004). ¿Quiénes deben procrear? Los médicos eugenistas bajo el signo social (México, 1931–1940). *Cuicuilco, 11*(31), 45–80.

Saavedra, A. M. (1934). *Eugenesia y medicina social.* Mexico, D.F., Mexico.

Sánchez-Bringas, Á. (2005). Prácticas reproductivas en el Distrito Federal a finales del siglo XX. En M. Torres (Ed.), *Nuevas maternidades y derechos reproductivos* (pp. 33–59). México: Colegio de México.

Santacruz-Varela, J. (2010). El aseguramiento de la salud en México y sus tendencias: Del mito al hito. *Revista CONAMED, 15*(4), 195–203.

Santillana, M. (1952, Octubre). Consideraciones sobre las calisficación etio-patogéncia de la infertilidad. *Estudios sobre Esterilidad, III*(4), 211. AMEE.

Sántos Haliscak, R., García Martínez, M., Galache Vega, P., Loret de Mola Gutiérrez, R., Morales Caballero, F., & Hernández Ayup, S. (1989). Ovum donation and gamete intrafallopian transfer (GIFT): A new reproduction concept. *Ginecología Y Obstetricia De México, 57*, 214–217.

Schurr, C. (2017). From biopolitics to bioeconomies: The art of (re-) producing white futures in Mexico's surrogacy market. *Environment and Planning D: Society and Space, 35*(2), 241–262. https://doi.org/10.1177/0263775816638851.

Segal, S. (1966, June). Family planning in national health programs. *Bulletin N.Y. Academy of Medicine, 42*(6), 447–453.

Serviere Zaragoza, C. F. (2003). Reproducción Asistida. Indicaciones. Sección XIII Reproducción Asistida. Cap. 39 (p. 285). In E. Vázquez Benítez (Ed.), *Medicina Reproductiva* (2º). México, D.F., México: Editorial Manual Moderno, AMMR.

Shalev, S., & Lemish, D. (2012). Dynamic infertility. *Feminist Media Studies, 12*(3), 371–388. https://doi.org/10.1080/14680777.2011.615627.

Sharp, L. A. (2000). The commodification of the body and its parts. *Annual Review of Anthropology, 29*, 287–328.

Shaw Kay, M. (2016). *Embodied agency and agentic bodies: Negotiating medicalization in Colombian Assisted Reproduction* (PhD Thesis). The University of Edinburgh, Edinburgh.

Shetty, A., Shetty, S., & Dsouza, O. (2014). Medical symbols in practice: Myths vs reality. *Journal of Clinical and Diagnostic, 8*(8), PC12–PC14. Research: JCDR. https://doi.org/10.7860/jcdr/2014/10029.4730.

Sierra García, F. (1964). Incremento demográfico y planeación familiar. *Estudios Sobre Esterilidad, 15*(3), 131–137.

Simpson, B. (2016). IVF in Sri Lanka: A concise history of regulatory impasse. *Reproductive Biomedicine & Society Online, 2,* 8–15. https://doi.org/10.1016/j.rbms.2016.02.003.

Soberón, G., Kumate, J., & Laguna, J. (compliladores). Cuahutemoc Valdés (coordinador). (1989). *La Salud en México: Testimonios 1988. Especialidades Médicas en México.* Tomo IV. Vol 1 y 2. Secretaría de Salud. INSP. El Colegio Nacional. FCE: Biblioteca de la Salud.

Sordo Noriega, A. (1951). Trascendencia de la Esterilidad. *Estudios Sobre Esterilidad, 3*(3), 113–115.

Soto Laveaga, G. (2005). Uncommon trajectories: Steroid hormones, Mexican peasants, and the search for a wild yam. *Studies in History and Philosophy of Science Part C: Studies in History and Philosophy of Biological and Biomedical Sciences, 36*(4), 743–760. https://doi.org/10.1016/j.shpsc.2005.09.007.

Soto Laveaga, G. (2007). "Let's become fewer": Soap operas, contraception, and nationalizing the Mexican family in an overpopulated world. *Sexuality Research and Social Policy, 4*(3), 19–33. https://doi.org/10.1525/srsp.2007.4.3.19.

Soto Laveaga, G. (2009). *Jungle laboratories: Mexican peasants, national projects, and the making of the pill.* Durham and London: Duke University Press.

Stephens, N., & Ruivenkamp, M. (2016). Promise and ontological ambiguity in the in vitro meat imagescape: From laboratory myotubes to the cultured burger. *Science as Culture, 25*(3), 327–355. https://doi.org/10.1080/09505431.2016.1171836.

Stern, A. (2002). Madres conscientes y niños normales: la eugencia y el nacionalimiso en México posrevolucionario, 1920–1940. In Laura Cházaro (Ed.), *Medicina, Ciencia y Sociedad en México, Siglo XIX.* Zamora: El Colegio de Michoacán, Universidad Michoacana de San Nicolás de Hidalgo.

Stern, A. M. (1999a). Buildings, boundaries, and blood: Medicalization and nation-building on the US-Mexico border, 1910-1930. *The Hispanic American Historical Review, 79*(1), 41–81.

Stern, A. M. (1999b). *Mestizophilia, biotypology, and eugenics in post-revolutionary Mexico: Towards a history of science and the state, 1920–1960* (Working Papers Series Center for Latin American Studies). The University of Chicago, 4.

Stern, A. M. (1999c). Responsible mothers and normal children: Eugenics, nationalism, and welfare in post-revolutionary Mexico, 1920–1940. *Journal of Historical Sociology, 12*(4), 369–397.

Stomer, M. R., Girault, P., & Wilson, L. E. (1984). *Combined Earth and Rock Bearing Foundation—Hospital Humana Mexico City D.F.* International

Conference on Case Histories in Geotechnical Engineering. 7. Retrieved form http://scholarsmine.mst.edu/icchge/1icchge/1icchge-theme1/7.

Stone, A. (1953a). Problemas de Fertilidad, Esterilidad y Población. *Estudios sobre Esterilidad, IV*(2), 66.

Stone, A. (1953b). Fertility problems in India. *Fertility and Sterility, 4*(3), 210–217. https://doi.org/10.1016/s0015-0282(16)31263-8.

Straw, C., Vargas, E., Viera Cherro, M., & Taminini, M. (2016). *Repdoçâo assistida e relações de gênero na américa latina.* Curitiba: Editora CRV.

Tamanini, M., & Tamanini Andrade, M. T. (2016). As Novas Tecnologias da reproduÇão humana, aspectos do cenário brasileiro, na voz e nas redes dos especialistas. In C. Straw, E. Vargas, M. Viera Cherro, & M. Tamanini (Eds.), *Rerpodução Assistida: e relações de gênero na América liana.* Curitiba, Brasil: Ed. CRV.

Tamez González, S., & Eibenschutz, C. (2008). El Seguro Popular de Salud en México: Pieza Clave de la Inequidad en Salud. *Revista de Salud Pública, 10*(Suppl. 1), 133–145.

Tamez González, S., & Valle Arcos, R. I. (2005). Desigualdad social y reforma neoliberal en salud. *Revista Mexicana de Sociología, 67*(2), 321–356.

Tamez, S., Eibenschutz, C., Camacho, I., & Hernandez, E. (n.d.). *Neoliberalismo y política sanitaria en México.* Foro Social Mundial de la Salud. Retrieved from http://medicinaweb.cloudapp.net/observatorio/docs/em/lg/EM2010_Lg_Tamez.pdf.

Tate, J. (2007). The good and bad women of telenovelas: How to tell them apart using a simple maternity test. *Studies in Latin American Popular Culture, 26,* 97–111.

Tatum, H. J., & Delgado-Garcia, R. (1968). Research on physiological aspects of reproduction. *The Milbank Memorial Fund Quarterly, 46*(3), 121. https://doi.org/10.2307/3349318.

Téllez Velasco, S. (2003). Presentación. In E. Vázquez Benítez (Ed.), *Medicina Reproductiva* (2º). México, D.F., Mexico: Editorial Manual Moderno, AMMR.

Tesauro, G. (1956). Conference programme. In *International Fertility Association.* Naples.

The First Test-Tube Baby Birth Watch in Britain for an Infant Conceived in the Laboratory. (1978). *Time, 112*(5), 58.

Thompson, C. (2005). *Making parents: The ontological choreography of reproductive technologies.* Cambridge, MA and London, UK: MIT Press.

Thompson, C. (2016). IVF global histories, USA: Between Rock and a marketplace. *Reproductive Biomedicine & Society Online, 2,* 128–135. https://doi.org/10.1016/j.rbms.2016.09.003.

Tlapanco Barba, R. (2003). Contraindicaciones para la reproducción asistida. Sección XIII Reproducción Asistida. Cap. 40 (p. 291). In E. Vázquez Benítez (Ed.), *Medicina Reproductiva* (2º). México, D.F., Mexico: Editorial Manual Moderno, AMMR.

Tompkins, P. (1950). A new journal. *Fertility and Sterility, 1*(1), 1–2. https://doi.org/10.1016/S0015-0282(16)30061-9.

Trouillot, M.-R. (1995). Chapter 1. The power in the story. In *Silencing the past: Power and the production of history* (pp. 1–30). Boston: Beacon Press Books.

Turner, F. C. (1974). *Responsible parenthood: The politics of Mexico's new population policies.* Washington, D.C.: American Enterprise Institute for Public Policy Research.

Unnithan-Kumar, M. (Ed.). (2004). *Reproductive agency, medicine and the state: Cultural transformations in childbearing.* Berghahn Books. Retrieved from http://www.jstor.org/stable/j.ctt1btbzpb.

Valdés la Vallina, F. (1954). *Orientación Internacional actual sobre esterilidad masculina, V*(1), 14–23.

Valdés la Vallina, F. (1964, Mayo–Agosto). Primera Reunión Anual de la AMEE Discurso del Presidente. *Estudios sobre Esterilidad, XV*(2), 56.

Valdés, L. M. (1980). Ensayo sobre política de población 1970–1980: Planificación familiar. *Estudios Demográficos y Urbanos, 14*(4), 467–480. https://doi.org/10.24201/edu.v14i04.471.

Vallarta-Vázquez, M. (2005). El consentimiento informado: un derecho reproductivo en México. In M. Torres (Ed.), *Nuevas maternidades y derechos reproductivos* (pp. 239–274). México: El Colegio de México.

Vázquez-Benítez, E. (Ed.). (1999). *Medicina Reproductiva en México* (Primera). México, D.F., Mexico: Manual Moderno JGH Editores S.A. de C.v.

Vázquez-Benítez, E. (Ed.). (2003). *Medicina Reproductiva* (2º). México, D.F., México: Editorial Manual Moderno, AMMR.

Vázquez-Benítez, E. (2008, Julio–Septiembre). La Asociación Mexicana de Medicina de la Reproducción 1949–2008. *Revista Mexicana de Medicina de la Reproducción, 1*(1), 3–14.

Vázquez, E., & Bruciaga, V. (1968, Septiembre–diciembre). Estudio de la Actividad Progestacional del Caproato de 19-Norprogesterona y sus posible aplicaciones en la fertilidad humana. *Estudios Sobre Esterilidad, XIX*(3), 128.

Velázquez Cornejo, G. (2012). Registro: Mexicano de Reproducción Asistida. *Revista Mexicana de Reproducción, 5*(1), 1–2.

Vera. (1986, December 13). El hospital, en venta, pero nadie compra. *Proceso.* Retrieved from https://www.proceso.com.mx/145031/el-hospital-enventa-pero-nadie-compra.

Viera Cherro, M. (2012). Inequidades múltiples y persistentes en el campo de la reproducción asistida. *Revista de Antropología Social, 21.* https://doi. org/10.5209/rev_raso.2012.v21.40058.

Viera Cherro, M. (2015). Sujetos y cuerpos asistidos Un análisis de la reproducción asistida en el Río de la Plata. *Civitas - Revista de Ciências Sociais, 15*(2), 350. https://doi.org/10.15448/1984-7289.2015.2.17157.

Villa Roiz, C. (1997). *Popocatepetl: mitos, ciencia y cultura: un cráter en el tiempo.* Mexico: Plaza y Valdés.

Wahlberg, A. (2016). The birth and routinization of IVF in China. *Reproductive Biomedicine & Society Online, 2,* 97–107. https://doi. org/10.1016/j.rbms.2016.09.002.

Wahlberg, A. (2018). *Good quality: The routinization of sperm banking in China.* Oakland: California University of California Press.

Wahlberg, A., & Rose, N. (2015). The governmentalization of living: Calculating global health. *Economy and Society, 44*(1), 60–90. https://doi. org/10.1080/03085147.2014.983830.

Walmsley, H., & Schurr, C. (2014). Reprodctive tourism booms on Mexico's Mayan Riviera. *International Medical Travel Journal, 3.*

Wang, J., & Sauer, M. V. (2006). In vitro fertilization (IVF): A review of 3 decades of clinical innovation and technological advancement. *Therapeutics and Clinical Risk Management, 2*(4), 355–364.

Weihe Edge, B. (2014). Infertility on E! Assisted reproductive technologies and reality television. *Feminist Media Studies, 14*(5), 873–876. https://doi.org/ 10.1080/14680777.2014.952875.

Weisman, A. (1953, Abril). La infertilidad marital debida a problemas emocionales no orgánicos. *Estudios sobre Esterilidad, IV*(2), 85.

Welti-Chanes, C. (2011). *La demografía en México, las etapas iniciales de su evolución y sus aportaciones al desarrollo nacional, 69,* 39.

Welti-Chanes, C. (2014). El Consejo Nacional de Población a 40 años de la institucionalización de una política explícita de población en México. *Papeles de Población, 20*(81), 25–58.

Wentzell, E. A. (2013). *Maturing masculinities: Aging, chronic illness, and Viagra in Mexico.* Durham: Duke University Press.

Wexler, K. (1995). Egg-swaping scandal still unfolding. *Washington Post,* 10.

Whittaker, A. (2016). From 'Mung Ming' to 'Baby Gammy': A local history of assisted reproduction in Thailand. *Reproductive Biomedicine & Society Online, 2,* 71–78. https://doi.org/10.1016/j.rbms.2016.05.005.

Winner, L. (1980). Do artifacts have politics? *Daedalus, 109,* 121–136.

Yovich, J. (1994). Transabdominal gamete intrafallopian transfer. In J. G. Grudzinskas, M. G. Chapman, T. Chard, & O. Djahanbakhch (Eds.), *The Fallopian Tube: Clinical and Surgical Aspects* (1st ed., pp. 213–227, Chap. 16). Springer-Verlag London Limited. Retrieved from http://www.researchgate. net/profile/John_Yovich/publication/235935827_Transabdominal_Gamete_ Intrafallopian_Transfer/links/02bfe5147c572c550f000000.pdf.

Zander-Fox, D. L., Henshaw, R., Hamilton, H., & Lane, M. (2012). Does obesity really matter? The impact of BMI on embryo quality and pregnancy outcomes after IVF in women aged ≤38 years. *Australian and New Zealand Journal of Obstetrics and Gynaecology, 52*(3), 270–276. https://doi. org/10.1111/j.1479-828X.2012.01453.x.

Zárate, A. (2015). Crónica acerca del Doctor Jorge Martínez Manautou, médico ilustre del IMSS. *Revista Médica del Instituto Mexicano del Seguro Social, 53*(2), 254–255.

Zárate, A., & Basurto-Acevedo, L. (2013). Notas históricas sobre la investigación científica en el IMSS. *Revista Médica del Instituto Mexicano del Seguro Social, 51*(6), 650–655.

Zárate, A., & Hernández-Valencia, M. (2014). El desarrollo de la endocrinología ginecobstétrica en el Instituto Mexicano del Seguro Social. *Perinatología y Reproducción Humana, 28*(3), 174–177.

Zárate, A., & MacGregor, C. (Eds.). (1987). *Manejo de la Pareja Estéri. Un libro para facilitar el tratamiento de la esterilidad.* México: Editorial Trllas.

Zárate, A., Canales, E., & MacGregor, C. (1976). *Esterilidad e Infertilidad.* México: Ediciones Científicas. La Prensa Médica Mexicana, S.A.

Zavala de Cosío, M. E. (1989). Fecundidad: dos momentos en la transición demográfica. *Demos, UNAM e-Journal* (2), 6–7. México, UNAM. www. ejournal.unam.mx.

Zavala de Cosío, M. E. (1992). *Cambios de fecundidad en México y políticas de población.* México: El Colegio de México / Fondo de Cultura Económica.

Zegers-Hochschild, F. (1996). *Registro Latinoamericano de Reproducción Asistida 1996.* Red Latinoamericana de Reproducción Asistida. Retrieved from http://www.redlara.com/images/arq/1996.pdf.

Zegers-Hochschild, F. (1998). Consenso latinoamericano en aspectos ético-legales relativos a las técnicas de reproducción asistida. Reñaca, Chile, 1995. Red Latinoamericana de Reproducción Asistida. Marzo, 1996. *Cadernos de Saúde Pública, 14*(suppl 1), S140–S146. https://doi.org/10.1590/ S0102-311X1998000500026.

Zegers-Hochschild, F. (2003). *Registro Latinoamericano de Reproducción Asistida 2003.* Chile: Red Latinoamericana de Reproducción Asistida.

Zegers-Hochschild, F. (2011). Barriers to conducting clinical research in reproductive medicine: Latin America. *Fertility and Sterility, 96*(4), 802–804. https://doi.org/10.1016/j.fertnstert.2011.08.043.

Zegers-Hochschild, F., & Galdames, V. (2001). *Registro Latinoamericano de Reproducción Asistida 2001* (p. 49). Chile: Red Latinoamericana de Reproducción Asistida.

Zegers-Hochschild, F., & Galdames, V. (2002). *Registro Latinoamericano de Reproducción Asistida 2002.* Chile: Red Latinoamericana de Reproducción Asistida.

Zegers-Hochschild, F., & Prado Aravena, J. (1990). *Registro Latinoamericano de Reproducción Asistida 1990* (pp. 1–13). Retrieved from International Working Group for Registers on Assisted Reproduction IWG website: http://www.redlara.com/PDF_RED/RLA-1990.pdf.

Zegers-Hochschild, F., Balmaceda, J. P., & Galdames, V. (2000). *Registro Latinoamericano de Reproducción Asistida 2000* (p. 45). Chile: Red Latinoamericana de Reproducción Asistida.

Zegers-Hochschild, F., Galdames, V., & Balmaceda, J. P. (1999). *Registro Latinoamericano de Reproducción Asistida 1999.* Chile: Red Latinoamericana de Reproducción Asistida.

Zegers-Hochschild, F., Galdames, V., & Schwarze, J. E. (2004). *Aniversario Registro Latinoamericano de Reproducción Asistida 2003–2004.* Chile: Red Latinoamericana de Reproducción Asistida.

Zegers-Hochschild, F., Galdames, V., & Schwarze, J. E. (2005). *Registro Latinoaméricano de Reproducción Asistida 2005* (p. 103). Chile: Red Latinoamericana de Reproducción Asistida.

Zegers-Hochschild, F., Galdames, V., & Schwarze, J. E. (2006). *Registro Latinoamericano de Reproducción Asistida 2006* (pp. 1–95). Chile: Red Latinoamericana de Reproducción Asistida.

Zegers-Hochschild, F., Galdames, V., & Schwarze, J. E. (2007). *Registro Latinoamericano de Reproducción Asistida 2007* (p. 75). Chile: Red Latinoamericana de Reproducción Asistida.

Zegers-Hochschild, F., Schwarze, J. E., & Galdames, V. (2008). *Registro Latinoamericano de Reproducción Asistida 2008* (p. 75). Chile: Red Latinoamericana de Reproducción Asistida.

Zegers-Hochschild, Schwarze, J. E., Musri, C., & Crosby, J. (2009). *Registro Latinoamericano de Reproducción Asistida 2009* (p. 78). Chile: Red Latinoamericana de Reproducción Asistida.

Zegers-Hochschild, F., Schwarze, J. E., Crosby, J. A., Musri, C., & Souza, M. do C. B. de. (2012). Assisted reproductive technologies in Latin America: The Latin American Registry, 2010. *JBRA Assisted Reproduction, 16,* 320–328.

Zegers-Hochschild, F., Schwarze, J. E., Crosby, J. A., Musri, C., & Souza, M. do C. B. de. (2013). Assited reproductive technologies (ART) in Latin America: The Latin American Registry, 2011. *JBRA Assisted Reproduction, 17*(4). https://doi.org/10.5935/1518-0557.20130062.

Zegers-Hochschild, F., Schwarze, J. E., Crosby, J. A., Musri, C., & Souza, M. do C. B. de. (2014). Assisted Reproductive Technologies (ART) in Latin America: The Latin American Registry, 2012. *JBRA Assisted Reproduction, 18*(4). https://doi.org/10.5935/1518-0557.20140018.

Zegers-Hochschild, F., Schwarze, J. E., Crosby, J. A., Musri, C., & Souza, M. do C. B. de. (2015). Assisted reproductive technologies in Latin America: The Latin American Registry, 2012. *Reproductive BioMedicine Online, 30*(1), 43–51. https://doi.org/10.1016/j.rbmo.2014.10.003.

Zegers-Hochschild, F., Schwarze, J. E., Crosby, J., Musri, C., & Urbina, M. T. (2017). Assisted reproduction techniques in Latin America: The Latin American Registry, 2014. *Reproductive BioMedicine Online, 35*(3), 287–295. https://doi.org/10.1016/j.rbmo.2017.05.021.

Zegers-Hochschild, F., Schwarze, J. E., Crosby, J. A., Musri, C., Urbina, M. T., & Latin American Network of Assisted Reproduction (REDLARA). (2016). Assisted reproductive techniques in Latin America: The Latin American Registry, 2013. *JBRA Assisted Reproduction, 20*(2). https://doi.org/10.5935/1518-0557.20160013.

Zhang, J., Liu, H., Luo, S., Chavez-Badiola, A., Liu, Z., Yang, M., … Huang, T. (2016). First live birth using human oocytes reconstituted by spindle nuclear transfer for mitochondrial DNA mutation causing Leigh syndrome. *Fertility and Sterility, 106*(3), e375–e376. https://doi.org/10.1016/j.fertnstert.2016.08.004.

Zhang, J., Liu, H., Luo, S., Lu, Z., Chávez-Badiola, A., Liu, Z., … Huang, T. (2017). Live birth derived from oocyte spindle transfer to prevent mitochondrial disease. *Reproductive BioMedicine Online, 34*(4), 361–368. https://doi.org/10.1016/j.rbmo.2017.01.013.

Index

© The Editor(s) (if applicable) and The Author(s) 2020
S. P. González-Santos, *A Portrait of Assisted Reproduction in Mexico*,
https://doi.org/10.1007/978-3-030-23041-8